WHAT EVERY ENGINEER
SHOULD KNOW ABOUT
CAREER
MANAGEMENT

WHAT EVERY ENGINEER SHOULD KNOW
A Series

Series Editor*

Phillip A. Laplante

Pennsylvania State University

*Founding Series Editor: **William H. Middendorf**

WHAT EVERY ENGINEER SHOULD KNOW ABOUT
CAREER MANAGEMENT

Mike Ficco

CRC Press
Taylor & Francis Group
Boca Raton   London   New York

CRC Press is an imprint of the
Taylor & Francis Group, an **informa** business

CRC Press
Taylor & Francis Group
6000 Broken Sound Parkway NW, Suite 300
Boca Raton, FL 33487-2742

© 2009 by Taylor & Francis Group, LLC
CRC Press is an imprint of Taylor & Francis Group, an Informa business

No claim to original U.S. Government works
Printed in the United States of America on acid-free paper
10 9 8 7 6 5 4 3 2 1

International Standard Book Number-13: 978-1-4200-7682-0 (Softcover)

Library of Congress Cataloging-in-Publication Data

Ficco, Michael.
    What every engineer should know about career management / Michael Ficco.
        p. cm. -- (What every engineer should know ; 43)
    Includes bibliographical references and index.
    ISBN 978-1-4200-7682-0 (alk. paper)
    1. Engineering--Vocational guidance. 2. Engineering--Management. I. Title. II. Series.

TA157.F48 2008
620.0023--dc22   2008013348

Visit the Taylor & Francis Web site at
http://www.taylorandfrancis.com

and the CRC Press Web site at
http://www.crcpress.com

Contents

II   Product Development

What Every Engineer Should Know: Series Statement

What every engineer should know amounts to a bewildering array of knowledge. Regardless of the areas of expertise, engineering intersects with all the fields that constitute modern enterprises. The engineer discovers soon after graduation that the range of subjects covered in the engineering curriculum omits many of the most important problems encountered in the line of daily practice—problems concerning new technology, business, law, and related technical fields.

With this series of concise, easy-to-understand volumes, every engineer now has within reach a compact set of primers on important subjects such as patents, contracts, software, business communication, management science, and risk analysis, as well as more specific topics such as embedded systems design. These are books that require only a lay knowledge to understand properly, and no engineer can afford to remain uninformed about the fields involved.

Preface

This book is written to fill the gaps in the background of certain engineers. These hypothetical engineers, because of education, experience, and general philosophical orientation, are very good at seeing some things and are completely oblivious to others. Unfortunately, a number of the things that escape their notice greatly affect their career advancement and even their day-to-day happiness. My intention is to give a context to the sometimes disconnected and confusing phenomena of the workplace society so that interested engineers can incorporate it into their particular reality. Once the subtleties of the workplace are recognizable in their view of the world, perceptive engineers can adapt their behavior patterns to be more successful and much happier in their chosen career.

This book is broken into two parts. Part I is specifically about the life and career advancement of the engineer. It starts in school and works its way to the experienced engineer, exploring various stops, diversions, and alternatives along the way. It also presents a view of the corporation as a living organism that has a unique personality that responds to stimuli of the world and evolves or perhaps dies.

Part II discusses engineering projects, product development, schedules, budgets, and related topics. This portion of the book is not about project management; rather, it is about the interaction of engineers and management working on projects in a corporate environment.

Author

Mike Ficco is a nationally recognized expert in embedded systems and embedded product development with nearly 30 years of experience in hardware, software, and systems engineering. His diverse background includes personal expertise and team leadership in the design, development, and implementation of military and commercial systems ranging from software products to satellite and terrestrial multimedia systems.

Mr. Ficco has worked in a senior capacity at extremely large corporations, was president of his own consulting company for 5 years, and participated in launching several venture-capital-backed start-up companies. In these capacities, he has invented new technologies, has led the development of complex systems, and has managed multimillion-dollar projects that productized hardware, software, digital, and radiofrequency devices.

Mr. Ficco has a B.S.E.E. and an M.S.E.E. from the University of Maryland and has completed the course work leading to a Ph.D. in computer science. He has attended a variety of industry training courses and has delivered presentations at major industry conferences. His design of a high-efficiency multimedia file system was published in the March 2003 issue of *Embedded Systems Magazine.*

Introduction

I'm an engineer. No, I'm a damn good engineer. I was born to be an engineer. I view things logically and believe nature to be knowable, predictable, and explainable. My favorite word is "how," and my second favorite word is "why." This is not a matter of training. It is just the way I am.

From my earliest memories I always wanted to know how things worked. In elementary school I began asking my parents for "geek" toys. In the days before personal computers I had a soldering iron, numerous electronics kits, and a "killer" chemistry set well before I entered high school. I also liked to read—especially comic books and science fiction. I loved building projects, many of them electronic. Few of my classmates read as much, and fewer still were making the chemical explosives, astronomical telescopes, model rockets, aircraft, walkie-talkies, and other electronic toys. I was self-entertaining and could spend hours, days, and even months in secluded pursuit of my latest interest. I gravitated to mathematics and science because they had patterns that I could recognize. Unlike the disorder of subjects like art and the arbitrary rules of English, math and science had "correct" answers. There was no subjective judgment. For me, math and science were very quick and efficient. There was little teacher judgment about the quality of my work. If the answer was correct, then it was correct.

In high school, counselors encouraged me to choose engineering as an occupation. Unfortunately, I knew no engineers and wasn't really sure what they did. As best as I could tell, they built cool stuff and had a lot of fun doing it. That was good enough for me. Soon it was off to college to become an electrical engineer. In college I learned that some math was really, really complicated. I learned of James Clerk Maxwell, who figured out how to explain a great amount of the world around us with four "simple" equations. I learned that 24 points out of 100 could be a "B" on an engineering test. I even learned that no matter how much you know there are always many more complicated topics waiting for discovery.

After graduating and working for a few years, I began to see repeating patterns. Many people simply did not think the right way to get complicated engineering devices to work reliably. A few people were just plain evil, and some others were self-promoting to the point of damaging projects. Nothing in my education had prepared me for the environment I encountered in the real world. It was composed of all different levels of skill and vastly different personalities, interests, and ambitions. It was much more complicated and diverse than I had expected. Indeed, the workplace reintroduced me to topics previously avoided: randomness, capriciousness, and sometimes outright deceit. It was fraught with the chaos of human behavior and personalities. In this environment I began the lifelong task of understanding motivations and goals of individuals and organizations. Some of the understanding was difficult to achieve because it involved understanding people who thought very differently from me. As creatures of finite consciousness we have little option other than to reference our experiences against our own personalities and backgrounds. I learned through some difficult lessons that not all of us think the same way or hold the same things to be important. I was quite surprised as I started to understand how very different some people were. In the workplace these differences sometimes resulted in superficial communications but left underlying serious misunderstandings. Sometimes these misunderstandings led to direct conflict with coworkers and with senior executives of the company. Such misunderstandings are not career enhancing and certainly do not help the project at hand.

As of this writing, I've worked for more than two dozen companies, and the previously described misunderstandings and communications problems were present to some extent at every one of them. Some personalities common to engineers are especially prone to these misunderstandings. These people are intelligent, ambitious, and hardworking but occasionally lack some key personality attributes needed for the successful navigation of the workplace society. It may take them years, if ever, to understand the complex and capricious rules of that environment. Some of the personalities around them are so very different from theirs that it is hard for them to understand the motivation of some of their coworkers, bosses, and corporate executives. Like me, many engineers enter the workplace wanting to "build cool stuff." Indeed, some of the better engineers I know became engineers for exactly that reason. Building cool stuff, however, doesn't always equate to getting the job done. Worse, many people in the workplace are driven by vastly different objectives. Sometimes the goals and objectives are so different that the engineer doesn't understand what his or her superiors really want. This lack of understanding may lead to direct and career-limiting conflicts with corporate executives. Understanding the corporate culture and workplace personalities is so important that some engineers may wish for an owner's manual for their career. They may want something to help them navigate the seemingly random whims of those around them who affect and sometimes control their lives. To them I offer this book. With luck, this book will not only help engineers and prospective engineers understand the workplace environment but will also allow "normal people" to better understand the engineers who occasionally think so differently from them.

I

The Engineering Career

Real knowledge is to know the extent of one's ignorance.

Confucius

1

Education

Introduction

In part, this book is intended to fill some of the holes I observed in my personal education. By the time I became a full-time member of the workforce, I had 22 years of schooling at various respected institutions. Despite the long years of study and learning, I found I was missing some fundamental skills. I could solve differential equations, but I had no idea how the loan company determined my monthly mortgage payment. I had no concept of escrow, mortgage insurance, or title searches. I was taught electromagnetic wave theory, but nobody ever told me how corporations worked, how executives were compensated, or where stock came from. I had π memorized to many decimal places, but I had no idea which silverware should be used for what purpose at an elegant dinner and, worse, had no idea why that might be important. As they say, the first step to being cured is to realize you have a problem. If you are a young engineer, chances are good that you have this kind of problem. You may be smart and you may be able to solve problems most people wouldn't even understand, but there are many things about the world you were never taught. Some of these involve the image you create. Others involve understanding the financial and social interactions that drive the world around you. Such understanding can have a direct effect on how successful you become. Once you graduate from engineering school, you must become a student of the world and continue learning.

The Early Years

Identifying future engineers is not terribly difficult. If you have a child who stares with wide-eyed wonder at how things work, she may become an engineer. If your child simply will not stop until he solves a problem, he may become an engineer. If children are comfortable not understanding something—anything—they will probably not become engineers. If children tend to operate freely on assumptions instead of facts, they are not likely to become engineers (at least not good ones). There are many reasons that children may become engineers, but my experience is that the best usually chose that profession because they simply love figuring out how things work and then making them better. Most engineers love designing and building things and would do so even if nobody paid them. When older, these people need enough money to lead a comfortable life, to buy a car, and to live in a nice house. They are often not interested in money for the sake of money. Rather, a desire for more money may be no more than a way to keep score as to who is the better engineer.

Emergence of Talent

All humans are an intersection of a broad range of skills at many different things. A good engineer can be tall and athletic or short and uncoordinated. Luck and statistics conspire to offer some people more options in life than others. Those who win the genetic lottery have many talents and therefore multiple career options. My experience with engineers is that the good ones share a few common traits. They are often inquisitive and don't willingly walk away from a problem before it is solved. They have an ability to focus on a problem for an extended period of time and have the self-confidence to attempt novel solutions. Intelligence is something of a bonus. Certainly a person of low intelligence would have a great deal of difficulty becoming an engineer, but curiosity and a dedication to solving the problem are far more important than brilliance. These attributes seem to be an inherent part of the psyche of engineers. I don't think you can really teach curiosity or a love of solving a complex problem. It's just the way you are, and it shows from early childhood.

Math and the Sciences

Children who become engineers frequently excel at math and science. Often this is not so much because they love these subjects or work hard at them but because math and science are relatively easy for them. Future engineers may not be the best spellers and their writing can be pretty bad, but you will repeatedly hear engineers say that the universe simply works the way they would expect. We could speculate that all human brains come prewired with some natural tendencies and expectations. If this is the case, we could further speculate that some percentage of humans' brains happen to be made to expect the universe to behave in a fashion that closely matches our current understanding of physical reality. Others might anticipate a world of magic and miracle solutions to problems and be comfortably able to believe impossible and mutually exclusive things. One would guess that the former group would have an advantage at being a good engineer whereas the latter should probably set their sights on a career in politics.

The anthropic principle states essentially that we would not be here to wonder about the universe unless the universe itself was conducive to our form of life. Perhaps we can define a cousin of the anthropic principle, the technopic principle, to state that engineers seem to be good at math and science because they would have difficulty becoming engineers if they were not. The important distinction is that being good at math and science does not make you a good engineer. Rather, being bad at them can prevent you from becoming an engineer. It is important to understand this distinction to comprehend that there is much more to being a good engineer than being good at math and science. Indeed, we are not even talking about intelligence or "being smart." Being smart helps you learn things faster, but having the right instincts and tendencies makes you a better engineer. The basis of being a good engineer is a deep-seated need to understand how things work. Everything else can be taught.

The Weeding Out Process

As children get older, they must have interest, talent, and dedication to become engineers. So much schooling and so many tests are required that the weeding out process is quite severe. Multitalented individuals who could have become engineers may get diverted anywhere along the way. Becoming an engineer may be too much work, or they could gravitate to a "sexier" profession. The geek stereotype of those who stay the course to become engineers has achieved truly legendary status. As a group, engineers are often though of as having thick glasses, not dressing very well, being shy or introverted, and perhaps having poor people skills. Unfortunately, there is some truth to this stereotype. Once, during lunch with eight coworkers, we observed that only one of us had perfect vision and nobody had on a tie. Of course, stereotypes are rarely accurate. All of us at the table, in our opinion, had exemplary interpersonal skills.

Anywhere along the way, multitalented individuals may get distracted by other professions that better fit their self-image, that have a higher salary, or that, for them, are more fun. An individual destined to become an engineer truly has fun solving problems. There is no substitute for the love of technology and the love of solving complicated technical problems. If someone just doesn't like the lifestyle of solitary dedication to linearly walking through a problem and figuring it out, no amount of money or pressure will make him or her good at it.

Educational Environment

School systems have a love–hate relationship with future engineers. They love them because they generally score well on tests and make the school system look good. They hate them because the school system has to accommodate a few pupils who stress the limits of the curriculum. Many high schools struggle to properly prepare a student for college engineering courses. They must find qualified teachers for a small number of students taking calculus, physics, and other advanced sciences. This can be expensive and appear to be elitist when a large block of students have difficulty meeting the minimum educational standards. Most school systems have finite resources and must make difficult decisions as to where to best apply them. It is hard to argue that 10 students need advanced calculus when 200 can't read very well.

Many reasons exist that make it difficult for both the schools and the young engineers in economically average or below-average communities. Wealthy school districts with highly paid instructors, largely professional parents, and skilled volunteers and assistants provide an extraordinary advantage not only to those who will become engineers but also to all college-bound individuals.

Social Interactions

Engineers tend to get paid more than many other occupations. Few engineers get rich, but most of them lead an economically secure life. Although this can be good for the adult engineer, it can cause problems for young students on this path. Everyone knows, or can reasonably assume, that many of the students who will become engineers, doctors, and lawyers will have a financially comfortable future. This can lead to some adults' subconsciously or purposely putting these "uppity" youngsters in their place with harsh treatment. Future engineers may have to overcome some amount of unfairness inflicted on them by teachers and other school administrators who resent the future prosperity the students' hard work and skill will bring.

There is also an undercurrent of technophobia in society as a whole. Most regular people do not have a good understanding of technology and therefore fear or distrust those who wield it. Many animals lash out at things they fear or distrust, and humans are no exception. Young people who will become engineers have not yet had the opportunity to acquire the affluent trappings of a good job, such as an expensive sports car, that can raise their social status. Without affluence and physical possessions to inspire respect and deflect hostility, they may have the added burden of dealing with those who scorn them because of their skills and studious work ethic. They may have to rise above considerable social denigration while studying the complex subjects of their curriculum.

Why do I say affluence inspires respect and deflects hostility? Well, it doesn't always. Sometimes it attracts a criminal element, and sometimes it actually provokes hostility through envy. However, being affluent certainly buys favorable treatment, influence, and latitude in actions and behavior. In addition, a substantial percentage of humans seem to be sycophants—or bootlicking, brownnosing, yes-men. Sycophants will be encountered throughout the life of an engineer. In high school these are the people who abuse the nerds in the hope that doing so will allow them to be accepted in more elite social groups. When they are older, the sycophants kowtow to the boss so they might get a promotion. They may also befriend and indulge quirky behavior of affluent people if they feel there is something to be gained. Sycophants would cruelly mistreat or bully these same people if not for their affluence or positions of influence and power. Wealth and power unquestionably buy deference from most people, and from this I correspondingly claim that, in large measure, affluence inspires respect and deflects hostility.

Bullies, sycophants, and brownnosers tend to view life differently from the more literal-minded engineers. Their motivation and goals are very different. For them life can be more about image, dominance, and power. For them life is not about solving cool problems but in getting ahead or at the very least preventing others from getting ahead of them. The earlier in life the engineer learns to detect and appreciate this fundamental difference in personalities and orientation the sooner they can adapt to the social complexities of the real world.

Free Time

Modern high-intensity and highly orchestrated high school curriculums are intended to help students get into good colleges. They challenge even the brightest with complex subjects and a great deal of work. The regular testing regimes of such curriculums necessitate many hours of study and benefit those students who are good at high-pressure, high-stakes testing. Clearly, those who do well in these systems are among the intellectual elite of our society and should do very well in college. There is, however, a negative side to this approach. While you are clearly selecting intelligent and hardworking individuals, there is more to being a good engineer. In fact, the high-pressure course load may actually be damaging the next generation of engineers by restricting their available free time—free time that would otherwise be used to build projects, to experiment, and to learn to design.

Free-ranging unsupervised technological exploration by high school students gives them a head start in learning to think critically in a step-by-step fashion. Before they get to engineering classes they will have found gaps in their knowledge and be wondering what a capacitor is for, how transistors switch, and what an operating system does. Some will even have answered these questions on their own before they get to college. They may have used their free time to build and program sophisticated robots or clever electronic kits or to hack into the national defense computer network. The important thing is they have actually done something more than read a book. These are the real engineers who will end up carrying projects and companies with their skill.

All colleges will teach the young engineer facts, but few will teach her the love of creation and the thinking process needed to actually assemble components and build a working project on her own. A young engineer who enters college knowing how to solder and how to program devices and write computer software will already have an important foundation on which to build. As facts are added, the engineer matures into a complete package with a thinking process that understands X and Y must happen before Z is possible. Those lacking such a foundation may miss important concepts and may suffer long-term damage to their high-level design and system architecture skills. It is unfortunate that some of this damage and corresponding disadvantage may be a direct result of spending years in an extremely aggressive and competitive educational system.

I don't know that lack of free time as a student damages future engineering skills, but I have repeatedly seen that many highly stressed and very busy engineers lose sight of the bigger picture and choose expedient and contrived solutions to appease the boss and to make the immediate problems go away—at a long-term cost to the project. One can only wonder if learning expediency habits in school will make this problem worse on the job. The engineering profession desperately needs individuals who instinctively understand the large number of components, tasks, subassemblies, and tests that must come together before a project can be successful. Such intimate understanding may be impaired if one's entire life has been more concerned with tomorrow's test than with making a fun project work elegantly.

Getting into a Good College

Those who choose not to participate or do not respond well to high-pressure and tightly planned high school curriculums may lack acceptable credentials to get into the highest-rated universities. How important is it to get your engineering degree from a "name-brand" university? Is it even a legitimate premise that there are good colleges and, therefore, bad ones? In my opinion, being a good engineer is more about a personality and a thinking process than about having a college degree. Having said that, however, getting a college diploma speaks positively about your work ethic and dedication to a goal. Getting the diploma from a noted university may speak a little louder.

However, the sheepskin may be saying more about the advantages with which you started life than about your hard work and dedication. A wealthy person will certainly have fewer obstacles getting into a top-rated university than a poor person. Likewise, a person who has to work to support his family may never have the opportunity to go to college at all.

Many universities offer an engineering education, but few are known around the world. There are many opinions on the advantages and desirability of getting a degree from one of the well-know universities instead of one of the more ordinary colleges. My personal view is there are two principal advantages to getting an engineering degree from certain well-known universities. First is name recognition. It does seem that hiring managers are more likely to invite candidates in for a personal interview if they have a degree from a well-known engineering school. Recognition of the university allows candidates to get their foot in the door and sell themselves.

The second advantage to attending a major engineering university is more a social advantage. While attending the university you will meet and interact with well-known teachers and important alumni and will be attending classes with future chief executive officers of corporations. The parents of students attending such expensive universities are also more likely to be affluent and therefore have friends and associates in positions of power. This provides the graduates of well-known engineering schools an extensive network of highly placed corporate leaders. These connections provide far greater opportunities than those available to, for example, a community college graduate.

Academic Achievements

It is necessary to get good, or perhaps even great, high school grades to get into a name-brand university, but how important is it to maintain a high grade point average in college? When I was in college people told me that companies did not want to hire students with really high grade point averages because the managers were afraid the person would be "too academic" and not practical enough. On the job as an employee of numerous companies I have found this is not the case. If anything, the opposite is true. That is, hiring managers appreciate high grade point averages. However, grades seem important only for the first job or two. After that your work experience becomes more important. I have not had my grade point average on my resume for many, many years.

However, honors programs and honor societies always have a place on the resume and provide a nice distinction from those lacking such accolades.

Graduate School

In theory, a person with a master's degree may command a higher starting salary than a person with only a bachelor's degree. However, this advantage quickly disappears, as subsequent jobs base a salary offer on your current salary, not your academic achievements. Graduate degrees can be useful throughout a career in competitive promotions. Some companies actually have an official policy that a position of director or higher requires a graduate degree. I'll note that this policy is often waived for favored candidates who happen to lack the necessary diploma. Nevertheless, a graduate degree can improve your chances and open doors in a competitive job market.

2

Framing the Corporate Landscape

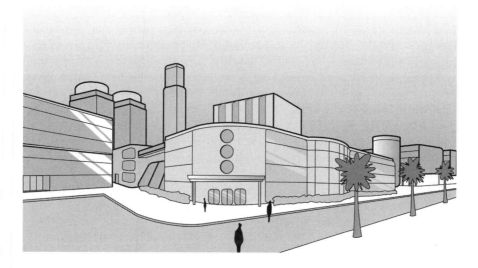

Introduction

Fresh out of school with my engineering degree I arrived in the workplace with little understanding of how corporations actually worked and no understanding of the many levels of bureaucratic hierarchy. Worse, I didn't know what I didn't know. To some extent this is a failure of the educational system, but much of the fault can be laid on my personality and upbringing. I'd experienced only two levels of bureaucracy in my entire previous life. There were teachers and students, and there were parents and children. In the workplace this translated to bosses and workers. It was difficult for me to comprehend the various levels of importance. I didn't really understand that some of my coworkers were more important than me. In fact, I didn't really understand that the bosses were more important than me. I had spent my life arguing with parents and teachers. Why wouldn't you argue with your boss? It took me a while to figure out that corporations are not a democracy and that most people don't care what the new kid thinks.

The First Weeks

Often new college hires arrive at their job, and their manager has no particular work for them. As a new hire, you might expect this. It is even understandable. First, your manager is very busy. Although you were hired because of your talent, it is unreasonable to expect a new college graduate to make a significant positive impact on a complex project that is running behind schedule. This means you probably can't help much right away. Your hiring is more of an investment in the future, and therefore interacting with you is a little lower priority than the daily crises that must be handled. If you are lucky, there will be a desk and computer ready for you when you show up at your new job. It is not uncommon for it to take a couple of weeks to get you a completely working system with account access, e-mail, and the other "necessities" of modern office life. It's not that you are being neglected. The reality is that everyone has bigger problems than doing things for you. Don't feel bad. It's not you. It's the workload.

Depending on the size of the company, you may be sent to some form of orientation class in your first days on the job. Such classes discuss the corporate culture, corporate holdings, and expected behavior as a member of and therefore representative of the company. With or without an orientation class you have government forms to fill out and medical insurance decisions to make. Some companies also have courses on the use of their e-mail system, software methodology, manufacturing processes, and so on. You may be given project documentation to read and might be taken around and introduced to the existing staff—or not. It depends on the personality, organizational skills, and to some extent the amount of free time of your manager.

The first couple of days will pass quickly as you try to remember where the restrooms are and everyone's name. After a while you might start going to meetings. You may be invited to a few; others may sound interesting, but you'll have to invite yourself. In some situations there is little guidance and hand-holding. In such cases, this is an opportunity for self-directed individuals to thrive. Lunchtime discussions, chatting with coworkers, and attending an occasional meeting can allow you to understand who is important and what is important to them.

"Listen much and talk little" is good advice for just about any situation, and starting a new job is no exception. Facts and technical correctness may not always be the most important things in the corporate world, so offering opinions can be risky until you learn the hot issues and the viewpoints of others in the organization.

It is probably not a good idea to complain to people that you have no work or are underutilized. This will very much embarrass your boss, whose job it is to keep you busy and productive. If you have not been assigned much work you have a golden opportunity to poke your nose into everything going on in the company and start offering to help on the things that most interest you. Going to your manager and asking approval may again embarrass her or you may be told no, limiting your chances to get involved in interesting work. Ask the engineers directly if there is something you can help on. It is best if you focus on helping engineers in your group. Helping engineers outside of your group may not be something your manager favors.

Corporate Organization and Operations

Corporations, even small ones, are composed of organizations with different responsibilities. The bigger the company, the more distinct are the groups, duties, and defined processes for the ways these groups interact. Different organizations in a corporation not only have different job functions but different behaviors, interactions, and expectations as well. Many companies, for example, discourage engineers from negotiating the costs of goods and services. This may be the job of the procurement organization even if they never heard of the component the engineer wants to buy, the company that makes it, or what a good price should be.

Corporations vary widely in their framework, and entire books have been written about corporate structure and the distribution of responsibilities. Any quick overview of groups most often encountered by a newly hired engineer would be a profound oversimplification and would have obvious omissions. The following is no exception and omits human resources, shipping-receiving, legal, and many more.

Business

This group consists of the people responsible for running the company. They make the final decision on which technologies deserve investment and which should be ignored. They also decide which marketing opportunities to pursue and which to abandon. Obviously, recommendations and information is received from all the other groups, but it is the business group that guides the overall strategy and activities of the company. As part of this job they hold press conferences and meet with stock analysts and court investors. If things go horribly wrong, they may face shareholder lawsuits and years of prison.

Engineering

For certain personalities this group is by far the most fun because you get to play with all kinds of toys and create new things. Engineers are responsible for designing and implementing new products as well as working with manufacturing to produce the products in volume. The overall engineering organization may be broken into hardware and software subgroups. In some companies this amounts to quite a clear division of labor, such as people in the hardware group who are capable of writing software rarely being allowed to do so, and vice versa. Other subdivisions of the engineering group may include test and verification, reliability engineering, and systems engineering. Systems engineering has the responsibility of ensuring that hardware and software work together in the overall product, or system, to meet the requirements. In the absence of a specific systems engineering group, the software folks take on some of this responsibility since their work tends to lie on top of the hardware and operate at a level closer to the customer's view of the product.

Financial

The financial folks keep track of the payroll, accounts receivable, accounts payable, corporate expenses, capital equipment, forecast earnings, and cash flow; they work with auditors and handle lots of government paperwork and requirements. Engineers work closely with the financial folks when budgeting and tracking project expenses. Engineering also works with the finance group when planning for additional staff or when economics force a reduction in staff size.

Manufacturing

This group is primarily responsible for building things or contracting out the building of things. The execution of this responsibility likely includes working closely with engineering to acquire materials needed to manufacture devices and to define and construct test plans and fixtures to ensure the devices are well made.

Marketing

Selling the current product line is often considered the domain of the sales group, and many companies have a distinct sales organization separate from marketing. However, sales and marketing both have a strong customer orientation and work closely together to create compelling sales and marketing campaigns for the company's products. This job entails interacting extensively with customers and prospective customers, attending conferences, and conducting seminars. All this human interaction at times necessitates heavy usage of the legendary marketing expense account. Sometimes engineers support the marketing and sales folks on a customer visit and get to partake of the expense account benefits.

Long-term survival of the company necessitates correctly anticipating future customer needs and demands. The marketing group's close ties with customers make it the obvious choice to define requirements for future products. One might assume that in a technical company this would be done by the engineers, as they are the experts in the technology and understand existing as well as near-future technical features that can be created with nominal effort. There is, however, some distrust of the engineers. On one side there is the fear that the engineers will avoid specifying requirements that will take a lot of work to implement. On the other side there is a fear that the engineers will specify geeky avant-garde features not of interest to the mainstream customers. The company has no such worries about the marketing group, whose lack of knowledge of the details involved in implementing the technology allows the marketers to specify features desired by customers without regard to the difficulty, time, or effort needed to make them work.

Occupational Safety

Not many people think of engineering as a dangerous profession, and, indeed, it is one of the safer. However, there are a few things to consider. By the nature of the work, engineers are exposed to far more microwave radiation than the average person. They also spend a significant amount of time in front of a computer monitor. They may from time to time be exposed to exotic alloys and compounds from circuit boards, wiring harnesses, and electronic components. There are, of course, the lunatic fringe and conspiracy theorists who are convinced that all such radiation and compounds cause tumors, cancer, and other horrible diseases. To my knowledge, no reputable studies have shown such claimed connections for environments normally encountered by typical engineers. Nevertheless, I would not be especially surprised if years down the road we determine that long-term exposure to low-level microwaves and such leads to an increased risk of cataracts and other soft-tissue damage. The point here is not that there are hidden health risks to engineers but that few if any professions have no health risk whatsoever. Gardeners and professional surfers risk skin damage from the sun, and old-time miners and chimney sweeps suffered from black lung. At least most engineers are well paid and spend the majority of their time in a comfortable environment.

Privacy

One might assume that an e-mail sent to your boss would be between you and your boss. This is not necessarily the case, especially for higher-level management. Executives often receive such a large volume of e-mail that their secretaries prefilter much of it. The secretary then highlights for the boss e-mails that need quick attention. One consequence of this is that secretaries to executives know a lot about what is going on in the company. This is generally okay, as the secretary is trusted to have this information. However, knowing that more than just your boss may be reading your e-mails should temper comments you convey in this fashion. Inadvertent dissemination of certain opinions can quickly generate extensive corporate intrigue, animosity, and embarrassment. An especially interesting situation may occur if another secretary helps out when the regular secretary is overworked or on vacation. Clearly, negative comments made about projects or individuals may get far more than the intended audience. Unintentionally wide distribution of viewpoints may also occur with voice mails and other forms of recorded communications.

Another feature of e-mails that occasionally causes problems is their long-term retention. Years after the fact, a court of law may subpoena e-mails to address any number of issues. Patent disputes, corporate squabbles, and discrimination claims may find reason to examine large numbers of historical e-mails. At times, statements taken out of context or made in jest may be quite damning and very costly to your company (or you). It may be beneficial to copy any e-mails associated with working around a patent or with avoiding some legal issue to the corporate attorney. Doing so would activate attorney–client privilege and perhaps shield incriminating comments from future courtroom eyes.

One need not send an e-mail to compromise privacy. Many office computer networks execute automated backup of selected directories or entire hard drives. This is accomplished by sharing the computer or laptop hard drive on the corporate network. Often this is done without the knowledge or explicit permission of the computer user. The employee, however, has often given implicit permission by signing the company's employment agreement. Irrespective of permission, system administrators may have access to just about everything on your computer. Stored employee reviews, letters of recommendation, or salary information might be freely available to the entire corporate information technology staff. Unless special precautions have been taken, this access may include all e-mail, corporate accounting, legal, medical, and business records. I've seen situations where company executives would have been horrified and apoplectic had they understood all the information freely available to certain low-level staff.

Privacy leaks do not occur only electronically. A large amount of sensitive data can be found lying for days at a time on fax machines and copiers.

Finally, some companies actually install "Trojan horse" programs on employees' computing equipment. This allows the corporation to track exactly what employees do during the day (or evening) with their corporate computer. Is this legal? Apparently so.

Corporate Culture

There are as many corporate cultures as there are corporations. Some emphasize hard work, some emphasize innovation, and some emphasize consensus building and being team players. Occasionally there are dramatic differences in the officially stated corporate culture's goals and their actualization. A company may repeatedly state that quality is of the utmost importance yet ship junk with impunity. A company may espouse the highest ethical standards yet bribe, cheat, and mislead. Words are just words, and sometimes the engineer is well served by direct observation of what is really important to and rewarded by the company.

Though the corporate culture will obviously have a relationship to the customers it serves, there is a strong and direct influence from powerful personalities at the top of the company. This influence can be simply stated as "what is important to my boss is important to me." In this way the corporate culture automatically reflects the personality and assumes the attributes emphasized by the leaders of the company. Sometimes this is good, and sometimes it is not so good. I'm familiar with a company in which the culture was dominated by hard work. While many companies emphasize the importance of dedication and long hours, this company was an extreme example. Every award praised long hours of labor, and every promotion extolled the commitment and effort of the promoted individual. The company wanted above all else a culture of hard work and that is exactly what they got—sort of. The company had successfully achieved a culture of appearing to work hard. Employees would get to work early in the morning and make sure they were seen. Many of them would then go off and read the newspaper or go shopping. Nearly everyone would leave voice mails early in the morning and late at night (or even in the middle of the night). Many of the voice mails were simple restatement of old information and added nothing to the resolution of problems. Worse, no labor-saving tools were ever purchased or built. Anything that reduced needed labor was counter to the very fabric of the company. It was this consequence of the culture that led to the ultimate demise of the business. Laziness can breed innovation, and the statement that all advancement originates with discontented individuals has a basis in truth. The corporate culture had driven dissent and innovation underground and trained employees to work on problems instead of solving them and moving on. The long-term effect was fatal to the company.

Large companies obviously have far more resources than smaller ones. This fact alone can result in vastly different corporate cultures. A small company may be a chaotic spectacle of ideas and work whereas many large companies have significant policies and processes in place that are intended to smooth workflow and to yield predictable and reproducible results. Most people will agree that while processes add predictability and reproducibility, they also add some amount of overhead. Small companies may not be able to afford to institute and manage large volumes of processes, but large corporations may fight a continuing battle against becoming too process heavy. A heavily process-oriented corporate culture is not necessarily a bad thing, especially for a manufacturing company. Occasionally processes advance beyond enforcing good design techniques and professional behavior to focusing on the ceremony and bureaucracy of the processes themselves. In these cases the large volumes of processes become overhead and an impediment.

Engineering processes also distribute responsibility and risk among multiple organizations and people. In a sense, engineering processes are like mutual funds. Mutual funds distribute the risk among numerous stocks. Processes ensure that many eyes look at a decision before approving it and that differing engineering personalities and talents adhere to a common methodology. Buying a single stock will most likely result in average stock performance, but in some cases you can either lose your money or hit a home run. Minimal or nonexistent processes can sometimes result in a brilliant engineer's skipping normal steps and making great intuitive leaps to achieve spectacular results in a short period of time. Like a single stock, however, the odds of such success are not good, and there is a corresponding risk of disaster. The odds become significantly worse if management is not adept at determining the capabilities of the engineers allowed to operate in a process-free environment.

Corporate cultures can evolve over the years as a company grows or changes. It can be very challenging to keep the culture positive and productive. Some companies are acutely aware of this and go to great lengths to ensure that employees fit a suitable mold. I'm familiar with several companies that administer a variety of forms of psychological testing of new-hire candidates and even existing employees. This investigation can range from simple word association tests to sophisticated and complex personality analysis. There are dangers to such profiling, not the least of which is that the company may end up getting exactly what it wishes. A corporate committee or executive, for example, may decide it is critical for all new hires to be team players. Overemphasizing the need to be team players may inadvertently filter out the more creative and aggressive team leader personalities. At first, corporate executives may enjoy the benefits of team players who follow direction without contention or dispute. Unfortunately, after a few years of enforcing such a policy, the company may find it is short on team leaders and innovators.

It is also possible that the intended psychological profiling simply doesn't work as desired. Some expensive psychological tests seem little more reliable than horoscopes dressed in pseudo-science and covered with innuendo and mysticism. The various sciences of psychological profiling get periodically discredited, and companies that sell such tests may provide only self-referential testimonials and anecdotal stories of success. The biggest problem with psychological profiling is the statistical variation in humans. It seems fairly easy to categorize the majority of people, but those who lie on the statistical fringes create enormous classification issues. Unfortunately, it is exactly the quirky and driven fringe personalities who often become the greatest leaders, scientists, and engineers. It may not be obvious that the profiling being used inadvertently discriminates against certain complex but extremely valuable personalities. Years may go by before it becomes obvious that the company's leadership and brainpower has been seriously impaired.

While corporate culture is subject to natural evolutionary pressures over the years, dramatic and significant changes can occur very quickly with personnel changes at high levels of the company. Most humans seem to have some amount of instinctive respect for authority. In many situations a boss, minister, or civic leader can issue a proclamation or make a decision, and few will challenge it or even think about it. Under normal circumstances, corporate culture and even society as a whole change slowly as issues get raised and traditional ways of doing things get challenged by those who think differently. Things can change quickly, however, when a new person enters a company at a senior level. Changes can be especially rapid and far-reaching if this new person has a strong personality and a specific agenda. It can be quite easy for a powerful person to get his or her way given the seemingly pervasive dual natural instincts of respecting authority and fear of falling out of favor and becoming an outcast. Dissent can be squelched any number of ways, ranging from questioning someone's patriotism or commitment to the company to outright dismissal. A new agenda and corresponding culture can rapidly fall into place in the absence of challenge or even open discussion concerning the wisdom of the new approach. Very quickly, anyone challenging the new order could do so only at the risk of his or her career.

Cultural changes are not always a bad thing, but it does seem there are many more bad ways to do things than good. As such, significant changes in an established and successful culture run a risk of being detrimental. The risk is gravely accentuated if the changes are accomplished under a veil of secrecy where open discussion and debate are discouraged. One would have to guess that any new agenda forced into place with coercion and intimidation and demeaning of dissenters should be a blinking and flashing red warning.

Power, Dominance Displays, and the Corporate Hierarchy

Numerous pundits have claimed that three things drive the world: money, sex, and power. Perhaps this is true, but my experience in the world of corporate engineering has not exposed me to much sex. The other two are ubiquitous. So many people jockeying for the next promotion and seeking power can become disruptive to genuine engineering and scientific activities. Occasionally, the more spectacular cases of politics' subverting the engineering method make headlines when something explodes or when people die. Although such well-publicized lapses are rare, less-dramatic blunders originate from misused authority and influence with alarming frequency.

Many years ago, when I was a college engineering student, a friend studying to become a medical doctor made an amazing claim. He said that instructors in medical school are deemed correct even if they say something that clearly contradicts the textbook. I countered that was not the case in engineering by far. If an engineering instructor stated something that conflicted with the textbook much of the class would identify his or her probable error and demand a clarification. In engineering, I asserted, facts are facts and are not subject to arbitrary revision. I can't speak to the truth of my long-ago friend's statement about medical school instructors, but I've been forced by experience to revise my opinion about engineering. I have seen firsthand numerous cases where facts were arbitrarily revised to meet the prevailing politically correct viewpoint. That portion of the class that would have challenged the professor never have graduated and made it into the engineering workplace where they would challenge such corruption of facts.

Most corporations—actually, most human activities—seem to operate very much on a caste system, where respecting the hierarchical chain of command is critical. The vast majority of people learn early that rocking the boat and challenging superiors leads to reprisal. Don't get me wrong; in general this is a good thing. However, the consequence of a large percentage of the population not following social direction is rather terrifying. On the other hand, continued scientific advancement requires challenging established views and considering alternate explanations for events. This need is well stated in the following quote from J. Robert Oppenheimer, the "father of the atom bomb":

> There must be no barriers for freedom of inquiry. There is no place for dogma in science. The scientist is free, and must be free to ask any question, to doubt any assertion, to seek for any evidence, to correct any errors.

For years it struck me as bizarre that some bad ideas were acted on while other good ideas were ignored. At first I rationalized this by thinking that the people running the company had access to much more information than I, so what seemed like a bad idea to me must really be good when the additional information is considered. Over the years, however, I observed many of the apparently bad ideas failing. Over the years I was also promoted a few times and gained access to high enough corporate levels to validate that, at times, management really did pursue bad ideas when better ones were readily available. What was going on? The people making the decisions were clearly (in most cases) intelligent.

I believe the problem was really my incorrect assumption that the various ideas were all given a fair consideration. It seemed to me that a good idea was a good idea and that good ideas were not restricted to any particular originating class. It was very foreign to my thinking for the quality and goodness of ideas to be defined by social rank. I didn't understand that many people are embarrassed when someone else comes up with a better idea—most especially if the person with the idea is of lower rank. My wife tells me that not being embarrassed when somebody has a better idea is a function of being secure and having good self-esteem. I don't know if this is true, but it took me a long time to understand that a junior person does not spring a random "good idea" on executives. Doing so invariably leads to the idea's being neglected or even wrongly discredited. In a way, this is understandable. The executives are very busy, and they have no particular reason to believe your idea is better than that of anybody else. This was made very clear to me on one occasion when a senior vice president of my company asked me in very frank terms why the company should follow my recommendations instead of those from more senior and presumably more knowledgeable people.

One reason to follow someone's recommendations is that they have established a track record of being correct. Unfortunately, it doesn't always work this way. Being correct in a fashion that supports an executive's inclination can be career advancing. However, disagreeing with a superior can result in your good advice's being ignored. Unwelcome counsel, even from someone with a long history of good judgment, can be rationalized away. Being proven correct later can be particularly damaging to a career. Certain personalities cannot brook this form of embarrassment. Offenders must be put in their place and perhaps discredited lest they become a political danger.

There is at times a dark side to the corporate chain of command. Just like any form of social hierarchy, those in charge feel the occasional need to demonstrate their authority and supremacy. Indeed, not all coworkers are kindred spirits. I have personally witnessed executives demeaning and insulting subordinates. I have seen an executive put his finger in the chest of a manager and explain in no uncertain terms that people were afraid of computers and that the Internet was a fad. I have seen a vice president extensively berate, for more than 5 minutes and in front of an entire staff, an underling for having dared to interrupt him. I have repeatedly seen competent and hardworking staff threatened with dismissal (as in, "If you can't do this I'll find somebody who can").

It seems that some percentage of the population simply enjoys bullying. Questioning why this is so seems pointless because there is no more explanation for this than why some people are shy and some are creative. A better question would be why highly capable employees put up with the bullying. It really does not bode well for society as a whole if some of the best educated and most intelligent people around don't vigorously oppose managerial bullying. While some engineers may have been teased in school, their knowledge and capabilities make them the "600-pound gorilla" in a technical company. It is absolutely alarming that these people tolerate abuse, unreasonable demands, and "motivational" threats of dismissal. If engineers can be intimidated in this fashion, what can be said of the remainder of society?

While only a small percentage of executives are bullies, most people who reach high levels of corporations are certainly very focused and driven. This focus and drive, however, can be subtly different from that of a similarly intense engineer. Motivated technical persons generally want to solve technical problems and in the process to make their project a success. Over an extended period of time this equates to a succession of varying goals as different projects and associated problems are encountered and overcome. Executives, however, remain focused on the distinct end goal of making themselves and their company successful. To certain executive minds, a successful project is only a stepping stone to greatness, not an end in itself. This difference in orientation and thinking can lead to a variety of misunderstandings. In these misunderstandings it is the engineer whose career is adversely affected so it is incumbent on him or her to understand and adapt to the mindset of the executive in charge.

Loyalty versus Ability

Corporate leaders are very busy people. They don't always have time to intimately understand every issue. They don't always have time to study new technologies and new ideas. They occasionally must rely on their instincts and trust their advisors. In certain circumstances this can translate to trust being the paramount attribute of an advisor. Trust, loyalty, commitment—all may become more valuable to the harried executive than accuracy and ability. With this insight one can understand why a loyal individual may be chosen over the brightest or most qualified to run an organization.

Inefficiencies that may occur as a result of such loyalty-based advancement can be viewed as inconsequential or manageable as the individual's loyalty ensures he or she will adequately follow direction. Problems may creep in, however, as this process trickles down layer after layer of the corporate hierarchy. In certain unfortunate situations, multiple levels of less than optimal people contribute to an intellectual downward spiral. In a worst-case scenario, lower-level management people become so inept that they can no longer identify good talent or good ideas and have no choice but to hire and promote exclusively by loyalty.

Chain of Trust

The perceived loyalty of an individual to his or her boss can dramatically affect access, influence, and advancement. Unfortunately, loyalty to your immediate boss is not necessarily loyalty to the king. The king's words may from time to time get corrupted as you transition through various levels of the management hierarchy. These words and judgments are heard by fallible humans, comprehended by finite intelligence, and repeated to the next level by those who may have specific and perhaps competing agendas. There is no outside divine force preserving the sanctity of the chief executive officer's (CEO's) intentions down through the corporate levels.

For many religions of the world there are holy and revered religious texts. Ardent believers are absolutely confident that these texts accurately capture the words and direction of God, a prophet, or an apostle. Few claim that God has spoken to them directly and told them of the inviolability of the sacred text. For all the others the only way to know that the revered text is genuine is because some trusted individual, or perhaps many trusted individuals, tell them so. How then did those trusted individuals come to know? The answer, of course, is that they too were told by trusted individuals. This chain of trust continues back in time until someone actually did speak directly with God, the prophet, or an apostle. For religious teachings, the chain of trust is divinely protected down through antiquity, guaranteeing the sanctity of the teachings in today's world.

Unfortunately, in the worldly affairs of man, direction from the CEO has no such deity-sponsored and -protected chain of trust. While some CEOs may think they are God, they are not, and their instructions and intentions can be misunderstood and corrupted on a regular basis. As an employee you really have little choice but to trust that your CEO's intentions are correctly represented by your immediate supervisor. While this corporate chain of trust can at times be very tenuous, you likely do not have direct access to your boss's boss's boss to validate the goodness of instructions you are receiving. This requires a little faith, and perhaps some prayers, that your instructions are legitimate and reflect the best interests of the company.

Keyhole Management

While the previously described chain of trust from the corporate leaders down to the rank-and-file employees may at times be tenuous, the path back up the chain of command can be nonexistent. Some managers rely exclusively on a few trusted advisors for information about the progress and activities of employees and projects. This approach is often termed *keyhole management* for the narrow window through which the leadership views events and activities happening in the company. The constriction of the keyhole can prove enormously efficient as it optimizes time and promotes careful focus on issues important to the executive. Unfortunately, delays, mismanagement, and other troubles may not be visible back up the management

chain because such information is filtered by the trusted advisors who are the keepers of the keyhole. While the filtering can lead to much happier executives, problems hiding in the darkness may fester and proliferate.

Such an isolationist management strategy may be the antipathy of the management style that walks around and mingles with all the workers and solicits their viewpoints. While both approaches have their benefits, neither is a panacea. "Management by walking around" can be inefficient as time is devoted to resolving the divergent opinions of lower-level individuals and perhaps discontented employees. This can distract from important issues and may result in ongoing criticism and continual second-guessing of decisions. Keyhole management, however, may result in a large number of activities' being concealed or misrepresented by the trusted advisors. While these critical individuals may be generally worthy of trust, history has repeatedly shown that gardeners rarely discuss the slimy things living under rocks in their own gardens. Sometimes the "gardeners" lack the skill to understand the significance of their problems, and sometimes for reasons of self-interest they intentionally hide the creepy crawly things under their supervision.

At times, the few trusted individuals with access to the next higher level of management continue to deliver a steady stream of glowing progress reports while they work feverishly to correct significant problems. It is commendable that these honorable individuals try to earn their pay by solving problems without burdening their boss. Sadly, the eventual result may be the boss's being surprised with catastrophic news about something he or she thought was going well. There are also times when a few trusted individuals reporting to an executive are not so worthy of trust. They may have observed that the bearer of bad news is often associated with and blamed for the bad news. Self-interest prevails over honor and righteousness, and these individuals avoid the stigma and risk of corporate excommunication by continuing to report good news to their bosses. The inaccuracy of the good news may escape detection for an extended period of time if none who know the truth are among those who actually speak to the executive.

One thing that has become painfully obvious during my career is that very few—very, very few—individuals interested in corporate advancement are comfortable with stepping forward and taking the blame for a problem or admitting to a significant mistake. This means that the efficiencies of keyhole management can be fully realized only in an environment that encourages and rewards independent supervision by those with dissenting opinions and competing agendas. There is good reason the fox is not allowed to watch the henhouse. Monitoring by people whose advancement and bonuses come from reporting the success of those being monitored is destined to fail, no matter how honorable the intention. People are people and most will eventually allow self-interest to dominate righteous intent. Even the most steadfast and virtuous may occasionally succumb to greed and hubris if there is little chance of exposure.

An organization that conducts its business in secrecy and squelches open discussion of issues promotes an atmosphere that can harbor corruption, stupidity, and incompetence for long periods of time. Only independent monitoring and oversight by those with contrary and competing agendas can reasonably be expected to ferret out idiocy, complacency, and inefficiency and expose them to the light of day.

Democracy

I used to think the miracle of America was capitalism, but as I gained job experience I began to understand capitalism was only part of the answer. Capitalism provides motivation for personal initiative and hard work, but this in itself isn't enough to explain the long-term success of the American experiment. Something else actually encourages creative and divergent thinking and the consideration of alternate approaches to problems. Something in American society, instead of stifling dissenting opinions and ideas, actually promotes them. That something is democracy.

Over time I saw the occasional negative consequences of keyhole management and began to understand that the miracle of America was really democracy and the division of power and responsibility across three branches of government with differing and somewhat competing responsibilities. This split in responsibilities assures, for example, that dissenting judges can differ with the opinion and direction of the president. Admittedly, judges could still be threatened or coerced into rubber-stamp acceptance of policy, but freedom always requires committed people and is never without risk. The free press can be viewed as a fourth branch of the government. In theory, it is independent of and beholden to none of the other three. The free press shares with the individual government branches the burden of reigning in judges, monitoring congress, or calling presidential direction to referendum.

It is important to understand that the benefits of democracy reach far beyond containing damage caused by a corrupt or incompetent judge, member of congress, or president. Well-intentioned, competent, and morally impeccable politicians also make mistakes. The different responsibilities of the branches of the American government ensure that the fox is not guarding the henhouse. The different responsibilities in a sense promote alternative agendas and impose checks and balances on each branch. Each branch has the responsibility to restrain bad intent or action by the others. The critical responsibility of the free press is to support this watchdog activity and to assist the three branches of government in calling attention to improprieties or errors in judgment. This creates a public forum for discussion and criticism. When issues surface and become sufficiently egregious, the average person is motivated to exercise his or her democratic empowerment and to remove offensive or inept people from government office. It is this democratic action that really limits how bad things can get before a course change is made, and it is the responsibility of the citizens to make use of available information and to hold civic leaders accountable for lies, misconduct, and foolishness.

Corporations are not a democracy. Corporations are a monarchy or perhaps an oligarchy. This can be incredibly efficient as it has been said that a benign monarchy is the most efficient form of government. There can be big problems, however, if the monarch is insane, evil, or incompetent. No amount of grassroots action or engineering recommendations can keep a corporation on a good course if the executives insist on a bad one. There is no corporate equivalent of the free press to poke its nose into projects and to expose irresponsible, incompetent, or malicious behavior. There is no democratic process for the staff to vote the CEO out of office, and there is no mechanism for the rank-and-file employees to hold senior management accountable for misconduct.

Democratic societies have accepted the need for secretive and classified government projects. The free press is prohibited from asking questions about them, and in many cases the government has decreed that citizens on the street cannot know anything about how their money is being spent. Management of these projects is occasionally by the very people who benefit from ever more grandiose and expensive clandestine projects. There is little "parental oversight" of the feasibility or benefit of the projects and generally no accountability for success or failure, as publicly admitting failure might be viewed as jeopardizing the security of the country. It seems that every year more things are deemed classified secrets and more money is spent on clandestine projects. One could easily believe this to be a bad situation if not for the commitment and honor of those involved.

In certain corporate cultures virtually every activity is the equivalent of such classified government projects. Information is not widely distributed, and there is every reason to be suspicious of those asking too many questions. In this corporate environment it is unimaginable that a manager would poke around in a peer's project. Corporate leaders can demand information, but those details might be conveyed only by the few trusted advisors and not those actually doing the work. Even the most trustworthy of advisors may occasionally allow self-interest to supersede truth and righteousness when there is little danger of lies and exaggeration being exposed. There is no investigative reporter to challenge assertions, and the corporate leaders have no reelection worries if they malign, belittle, or undermine their staffs.

Insecure or megalomaniacal CEOs may gradually oust opponents and consolidate more and more power. This can happen quite easily because there is no corporate law protecting free speech of the employees. There is no corporate equivalent of the Bill of Rights to protect jobs of employees who speak out against the directives of corporate management.

The First Amendment says:

> Congress shall make no law respecting an establishment of religion, or prohibiting the free exercise thereof; or abridging the freedom of speech, or of the press; or the right of the people peaceably to assemble, and to petition the Government for a redress of grievances.

Amendment I, United States Constitution

Rule, or Die Trying

There is no guarantee of free speech in a corporation. You are paid at the will of the corporate leaders and you keep your job only so long as they wish to retain you. Criticizing the corporate leaders may result in termination but can also result in subtle and indirect reprisals. The company can exert enormous financial pressures on an outspoken critic. In some corporations, an individual speaking negatively of the leadership may receive no overt indication of displeasure but might never get another good raise or quality work assignment. Being an outstanding engineer can buy some leniency, but ultimately certain executive personalities will readily sacrifice the good of the corporation rather than tolerate criticism. Certain personalities were born to rule, and they will stop at nothing until they are in charge and giving the orders. This personality attribute can be associated with a competent intelligent person, but it can also belong to an incompetent idiot. Unfortunately, being an idiot does not lessen the drive of these persons to issue orders.

Random chance assures that unskilled individuals obsessed with being in charge and giving orders will occasionally rises to some level of corporate power. Given a group of such people, some by pure dumb luck will preside over projects that become successful despite bad leadership. Those people have a chance for further advancement. Over time, some of them may reach very senior levels of a corporation. Here we have constructed a plausible scenario where corporate leadership can occasionally be seized by someone so incredibly driven to rule that he or she will denigrate coworkers or destroy the company rather than admit an error or fall from leadership. While being intelligent and capable may make it easier for such a driven person to reach the top of the corporate hierarchy, being incompetent does not preclude it. The message here is clear. Engineers may find it advantageous to avoid offering suggestions that reflect negatively on the corporate leadership until they understand the personalities involved. It is important to recognize that there is always some risk associated with telling the person who signs your paycheck that he or she is wrong.

In a corporate setting the "rule or die" personality can, and sometimes does, kill the company. Various circumstances and corporate cultures can result in highly controlled engineering staffs that, like highly controlled societies, suffer a loss of innovation, joy, and experience a general intellectual stagnation. This can be crippling and terminal for companies because it makes them an easy target for more enlightened competitors.

Enlightenment and Reason

To me it seems there is remarkable relevance to analogies and comparisons between corporations and society as a whole. One could view some companies as medieval feudal societies where a monarch and lords preside over the workers. In a few companies, this more closely resembles the Dark Ages, where rulers issued proclamations, dispensed directives, discouraged challenging the status quo, encouraged the corporate religion, and granted indulgences to the faithful. In many cases the leadership genuinely fails to see itself in such a negative light. The perception might be of only providing guidance and setting expectations for the employees. The real burden on the technical staff, and therefore on the long-term success of the company, need not be bad direction or incompetent management. Discouraging critical thinking and the consideration of alternate approaches can damage a company even when accompanied by capable direction. Blind acceptance of direction is risky in many of life's endeavors, and an engineering staff with this behavior may under-perform comparable raw talent that is repeatedly challenged to ask questions and innovate.

Innovation and progress require thinking outside the box. Innovation inherently means disregarding the established or traditional way of doing things. At times, new ways of doing things may inadvertently be a step backward, but continuing with the traditional approach guarantees eventual stagnation. Some corporate leaders exercise a totalitarian management style that brooks no dissent. Intellectual dissent and new ideas and approaches can be terrifying for the leader who must rule and dominate and cannot endure the shining of other corporate stars. Encouraging contradiction or debate is far removed from their realm of thinking. They give the orders and expect them to be followed.

Leaders dispensing absolute directives, discouraging critical thinking, and actively suppressing the investigation of competing ideas smells remarkably like the Dark Ages. During this time of intellectual stagnation misguided clerics stated with absolute certainty that the earth did not move and taught that disease could be cured through prayer. Instigators of change and proponents of other beliefs were viewed as heretics and disruptive people who were subverting proper society. For hundreds of years these troublesome people were dealt with harshly. By some accounts the intellectual suppression of the Dark Ages only succumbed due to indiscriminate killing by the Black Plague. Death was dealt no matter the station in life, the goodness of one's existence, or how fervently one prayed. Those issuing proclamations, warning of damnation, and fervently praying died alongside common thieves. Eventually some began to believe that perhaps leading a good life, worshipping God, and praying wasn't enough. Some began to believe that God wanted them to think, investigate, analyze, and make use of their God-given ability to reason. Over time, a critical mass of such believers was able to transform the Dark Ages into the Age of Enlightenment. The stranglehold of the clerics claiming to represent the will of God was broken, and the Renaissance was born.

In a corporation run like a totalitarian regime, the Black Plague corresponds to the competition. Competition will kill many companies. Some will grasp the danger presented by the competitor and understand that they must adapt or die. To adapt, they must loosen the controls that have held innovation in check. They must consider new ways of doing things, and they must unleash the repressed creativity of their technical staff. Unfortunately, some companies learn too late that they have excommunicated and driven away the innovators, unconventional thinkers, and "troublemakers" who might have saved the company. They moved to a freer society where their rebellious ways would be appreciated or at least tolerated. It is not surprising to me that great countries and great companies grow out of the hard work of people who encourage reflection, discussion, and debate and who allow those with dissenting opinions to say their piece. Progress accelerates when people argue among themselves in search of the right approach. Great things happen when those in charge allow or even encourage thoughtful criticism. Criticism of bad laws, bad designs, and bad thinking is necessary to bring about progress. Progress slows when the "right answer" is dictated from above and disagreement is not tolerated. This can be true even when the dictated right answer is indeed the right answer.

Irreverence, Malcontents, and Progress

Some managers, kings, presidents of nations, and corporate executives fear the unpredictability of the innovative personality. Some creative or deeply analytical people don't always play by the rules and follow the chain of command. They may see fault with the prevailing thinking, and they might criticize leadership or leak incriminating documents. A number of them are just not team players and at times can't be trusted to do what they are told. Unfortunately, all of this is true. What makes it worse is that we also owe just about everything we have to these outcasts who were not happy with existing social conditions, the belief systems of the masses, limitations of technology, the capabilities of musical instruments, ways of painting and sculpting, and everything else the human mind has invented and improved.

Many ideas viewed as dangerous and socially disruptive when first presented have since taken hold. Ideas as fundamental as suggesting we not use other humans as slaves or that the earth flies through space in orbit around the sun were not well received when first introduced. The history of mankind has been a continuing struggle between those wishing to control and suppress and those who want to explore, analyze, and push back the darkness of the universe. This is true of corporations and society as a whole. People are people whether they are managing a company or ruling a country. The suppression of the day might be religion run amok, onerous copy protection, McCarthyism, proprietary hardware, Stalinism, banning specific types of research, blocking radio broadcasts, or prohibiting the dissemination of scientific knowledge. Complicating matters is that those wanting to control or suppress often think they are doing it for moral and ethical reasons. They want to help misguided folks or to prevent good people from being seduced by bad ideas. Unfortunately, competent people who are intellectually and morally committed to a cause or approach can sometimes be wrong. The only protection is honest discussion and debate by openminded people willing to listen to reason and consider factual evidence. The Danish physicist Niels Bohr said this well: "The best weapon of a dictatorship is secrecy, but the best weapon of a democracy should be the weapon of openness."

Sometimes the suppression is social, sometimes it is technical, and sometimes it is both. The Nazi war machine produced some amazing technical achievements, but their racial intolerance was historically ruthless. The famously democratic and philosophical ancient Greeks achieved a level of technical enlightenment not replicated for a thousand years. Medieval Europe suffered centuries of intellectual and cultural stagnation while in the control of oppressive religious zealots. Examples are legion for both corporations and countries. The evidence is clear that tolerance of new ideas and divergent viewpoints generally leads to explosive growth in the areas of tolerance. The evidence is also clear in both nations and corporations that tolerance, innovation, and rapid advancement in one area may be accompanied by suppression, discrimination, and stagnation in another. Success or failure, freedom or censorship—these are functions of the interests, knowledge, and policies of the corporate or national leaders.

The executive giving the orders may be brilliant, but nobody—nobody—is more brilliant than the assembled mass of all humanity. A gifted CEO channels and guides creativity rather than smothering it and demanding things be done his or her way. A secure and genuine leader challenges subordinates to surpass the master. He or she does not assign penance to those who trespass against their masters' beliefs and direction. Anarchy is bad, but challenge, debate, and discussion are to be treasured. The history of corporations and nations proclaim that freedom to discuss, debate, and challenge results in the explosive growth of ideas and technology whose benefit reaches far beyond the original inventors. Each fundamental innovation morphs and inspires derivative advancements, creating jobs and raising the overall standard of living.

Unfortunately, innovation can sometimes be feared. At times, entrenched interests benefiting from the status quo may try to crush the new ideas that threaten their traditional way of life. These entrenched interests may want nothing more than preservation of their comfortable lives, but they might also be quite malicious and evil. The perpetuation of their riches and comfortable lifestyle may be contrary to the interests of their company, country, or humanity as a whole. The pursuit of their nefarious plans may involve the manipulation of altruistic and honorable people whose critical thinking skills are not well developed. These devious individuals endeavor to keep themselves in power by fabricating a conspiracy that presents the threatening idea or technology as un-American, godless, risky, money losing, sexually perverted, or otherwise evil. Of course, the proponents of such an idea must therefore be sociopaths, heathens, or misguided individuals who must be stopped at all costs.

The well-chosen audience will be of good and moral people who by personality or upbringing are poorly prepared to doubt or criticize authority figures purposefully lying and manipulating them. Successfully persuading a meaningful number of the target audience that the cause is righteous is a double win for the manipulators. Numerous followers of obviously high moral character seemingly validate the goodness of the cause championed by the entrenched interests. This creates a positive feedback loop where more people join the cause, adding ever more credibility. This manipulation of good but naive people into directed beliefs is an unfortunate fact of life in modern democracy.

There are also times when the manipulators are far from evil. They may be extraordinarily moral individuals who want nothing more than to "save" folks from what they perceive to be an evil path. Unfortunately, history has shown that paths sometimes considered immoral were actually correct. The Copernican theory, which held that the Earth is one of several planets revolving around the sun once a year and turning on its axis once a day, is one example of this. In the future, global warming and evolution may, or may not, be in the same category. The point is, we just don't know, and only factual research will tell us the truth. Firmly believing something does not always make it true. Certain personalities long for the security blanket of well-established beliefs and fear the opportunity and responsibility of creating new cultures, customs, or processes. It is often so very much easier to continue doing things the old way that a new way is cursed and discredited. Rev. Martin Luther King Jr. cautioned against this impediment to progress: "The soft-minded man always fears change. He feels security in the status quo, and he has an almost morbid fear of the new. For him, the greatest pain is the pain of a new idea."

Bureaucracy and regulations can be powerful enemies of advancement as old ways of doing things and making money desperately cling to the world they once knew. Companies used to making money in a certain way can be monstrously lacking in interest or acceptance of changing their way of life. They often continue eating at the old food trough until all the food is gone. They desperately seek to outlaw the new threatening technology and to banish or jail the disruptive proponents of the new order.

The most explosive growth of technology occurs in the Wild West mode before the sheriff and laws move in to protect intellectual property and ensure orderly progress. Unfortunately, orderly progress can be the antithesis of fundamental advancements that spur an avalanche of new ideas and products. Throughout history the introduction of certain technologies has revolutionized the world, altered society (perhaps not for the best), created mammoth corporations, and destroyed entire industries. Prior to the phonograph, musical performers often required affluent patrons to pay for or otherwise provide the necessities of life. Sales of records, tapes, CDs, and royalties from broadcasts of artists' works have since created large numbers of wealthy performers and bureaucrats. Internet piracy now threatens a way of life and a business model that didn't even exist 100 years ago. Somebody usually makes money off a revolutionary new technology. Conflicts occur when the newly rich wrestle power and control away from the old rich.

The telephone, the automobile, microprocessors, personal computers, the Internet, fire, the wheel, and many other innovations changed the world in ways the early tinkers could never envision. No corporate 5-year plan can ever plan for worldwide technological revolution. Maybe it's not a good idea to dump sewage into the river and spew smoke into the air. Perhaps investment is needed in a new technology disparaged by the CEO. Maybe the central processing unit (CPU) chosen by the system architect is not powerful enough. Challenging the status quo or proclamations of important people can be very difficult. Introducing new ideas can require a mixture of great self-confidence, disregard for personal benefit, and sometimes disregard for personal safety. In extreme cases, new ideas must question hundreds or thousands of years of tradition and teaching. Consider the incredible arrogance of saying, "The entire human race is wrong and I am right." Yet sometimes the arrogance is justified and the new way is a great advancement. A remarkable number of famous scientific innovators were highly irreverent. Galileo was well known for his independent thought and considerable arrogance, and Albert Einstein's distaste for rote learning highly irritated his teachers. Einstein regularly rebelled against authority, and his independent streak provoked animosity in his university professors. In a moment of reflection an elder and famous Einstein once said, "To punish me for my contempt for authority, fate made me an authority myself."

Unfortunately, it is precisely the irreverent, eccentric, intellectually arrogant, and at times abrasive personalities who risk disgrace or death to drag the human race kicking and screaming into the future. Extraordinary personalities who dared risk excommunication, torture, and public humiliation brought us the solar system, revolted against colonialism, and made us aware of racial discrimination. Those who suggest the directives of a corporate executive are flawed or denigrate the designs of the senior system architect risk subtle reprisal or outright dismissal. However, all progress depends on the passionate malcontents whose beliefs are so strong that they challenge accepted dogma—consequences and reprisals be damned.

It must be this way, for those who see life and problems in a traditional and respectful way simply cannot reach beyond their ordered and comfortable world to solve the difficult problems. Those who let others tell them how to think or who follow directions without questioning are incapable of inventing breakthrough technology. New ideas and approaches can be disruptive and painful to organized, harmonious, and conventional ways of thinking. There is a natural antagonism between new and impertinent points of view and the bureaucratic establishment. From the perspective of many corporate executives, these passionate malcontents are impossible to deal with. They just don't listen, and there is no telling what they will do.

The challenge for such innovative thinkers is to mask their true nature long enough to reach positions of power within their corporation. They must smile and thank the boss for the suggestion when it is the most stupid thing they've ever heard. They must avoid confrontation with their superiors and subtly exert creativity on their projects to avoid alarming the corporate leadership with controversial or heretical approaches. Intelligence, creativity, and commitment can accomplish nothing if the possessor never achieves a sufficiently responsible position to make good use of his or her abilities.

3

On the Job

Introduction

If you are a lucky person you will instinctively know how to behave in a career-enhancing fashion. Unfortunately, too many of us stumble and fall when confronted with complex professional situations. The economic consequences of such missteps can be severe, and your quality of life, happiness, and personal self-image may suffer badly from faux pas that might have been avoided. Sometimes learning is painful or comes only after a great amount of introspection, and sometimes situations and people take on highly visible or extreme traits. They become almost caricatures. In these cases, an incident or individual stands out from the background clutter of life and comes into clear focus. Suddenly you see for the first time a life template that allows previously encountered events or people to fit into a model about which you can think and speak. You become newly aware of connecting relationships between your words or conduct and resulting behavior of those around you. Be vigilant for these life templates and the learning experiences they can offer. They can be immensely helpful in improving your response to events and your interaction with those around you. Better social interaction with coworkers can dramatically minimize damaging blunders and can result in a more successful and satisfying career.

Naturally good technological or social instincts are just that and have little or no dependency on the career level or job assignments of the person possessing them. For many of us, however, lessons often need to be learned the hard way. If you are fortunate, gracious people will tell you what you are doing wrong. Hear their words, assess the accuracy of what they are saying, and act on them. Having said that, I will add that exhibiting this type of learning behavior as a senior person may make you seem more genuine and approachable, but it may also minimize your authority and dominance over your employees. To some, a senior person who takes advice from underlings may not seem to be in charge and leading the organization. This is an image issue and is independent of whether the advice was good and the learning was beneficial. For some personalities, this more humane, approachable, and fallible persona is anathema that can never be tolerated no matter what the consequences. Others, perhaps those more intrinsically secure, are willing to hear criticism from subordinates when it is offered in a constructive fashion. As a junior person you may benefit by withholding your opinions and advice until you understand how the recipient will react.

The Role of Experience

As noted already, naturally good instincts are, well, natural. They are a gift of genetic accident or perhaps luck of upbringing. Experience, however, is not a birthright but is available to everyone. On the job, experience is something everyone goes through, but there are as many different kinds of experience as there are people and jobs. You can have experience researching fundamental technologies that may take years to become practical, or you may help develop a product that makes your company billions. Some circumstances allow you to better understand and navigate the corporate political landscape, and some allow you to become a more proficient technical person. Interestingly, experience does not seem to play a big role in some aspects of engineering. Experience seems to hone or refine existing skills rather than to create a new lifestyle. It appears that beginners who lack creativity, jump to conclusions, or carelessly ignore defects continue to do so after many years on the job. Experience, in short, does not seem to be able to make a bad engineer into a good one. A friend of mine said it well: "An idiot with 20 years' experience is only an experienced idiot."

Experience teaches what problems are likely to occur and provides a frame of reference for methodologies that may allow rapid resolution or perhaps even complete avoidance of problems. The engineer may learn that signals with fast transitions can cause noise, but he can also learn that telling the boss that he or she is wrong can lead to reprisal or other negative consequences. The great advantage of experience is that it allows the engineer to better anticipate and respond to issues with a previously learned and rehearsed bag of tricks.

Independent of such practical issues of skill and ability, political realities of the workplace may not allow assignment of much responsibility to junior people. On a major project a manager coworker of mine assigned his best person to several of the most critical tasks. Unfortunately, his best person was a recent college graduate of significant skill. The schedule was rejected when it was presented to senior management. One of the executives was blunt: "I can't accept a schedule with these critical tasks assigned to such a junior person." The schedule was appropriately revised. The new version assigned a senior engineer to the tasks. The reality-check footnote was that the senior engineer merely assisted the junior engineer in completing the tasks. Although the senior engineer was not personally able to perform the work, his name and experience level were politically acceptable.

In more tolerant environments you may be allowed to assign a lower-level person to an important task, but there can still be political consequences. If something goes wrong you can be criticized for inattention to the need of properly staffing the project with the right people. Conversely, as the actual talent level of engineers is relatively invisible to the layperson, assigning a senior person of lower intrinsic skill can buy a political free pass should disaster occur. With the assignment of a senior engineer, it's clear you did everything you could to assure the success of the project.

Gaining seniority has several advantages. First, you make more money. More experience also makes you eligible for assignment to harder problems and greater responsibility. It may also force you to make fundamental decisions about your career. You may have to decide whether to stay technical, move into engineering or project management, or go over to the dark side and move to marketing. At times, you must decide to stay at the same job or move on. All the education, all the commitment and hard work, all the decisions really come down to determining what works for you. Unfortunately, this issue can become complicated because what works for you may vary throughout your career. One year the most important thing in the world might be being "successful." At some other juncture you may be driven to give orders and run the show. There could be times when you can't imagine anything more important than playing with the technical toys.

The world of technology changes quickly, and it can be difficult to return to hands-on technical work after a few years as a manager. On the other hand, a working technical person can feel frustrated at not having the authority to run a project the "right way." Career missteps and poor decisions can result in an engineer's feeling trapped in a hated job, believing he or she has lost or never developed the skills needed to land a desired position. Even if no missteps are made you might become trapped by your own aptitude in a particular field. You could become typecast or pigeonholed as the person who does "X." From the corporate perspective, always assigning the same knowledgeable person to do the work can be very seductive. There is no overhead of learning and no need for time-consuming knowledge transfer. The work just gets done. Such typecasting may be expedient but can result in poor distribution of knowledge important to the company. Just as it can trap a person in a pigeonhole, it can create prima donnas who make demands and hold the company hostage. Nevertheless, some companies are so bad at such typecasting that breaking out of it may require changing jobs. Even this is not always easy as it may be necessary to overcome the stereotype-driven preconceived notions of the hiring manager for the new job.

An engineer may leave a company because he was pigeonholed or for any number of reasons. Engineering organizations expect a certain amount of job turnover, but an excessive amount of churn becomes disruptive. When too many engineers leave in a short time, it damages the tribal knowledge of a company. No matter how well projects are documented, some of the problems solved and some of the learning exist only in the heads of the engineers involved in the work. This undocumented reservoir constitutes the company's tribal knowledge. Excessive turnover disrupts the dissemination of this knowledge and negatively impacts the capabilities of the company.

Excessive departures aside, one of the real tragedies of corporate engineering is the conversion of a good and experienced engineer into a bad manager. The workplace is littered with good engineers who loved their job but became bad managers to make more money. Of course they don't intend to become bad managers, but the odds aren't good that these people would excel at both jobs. Worse, their heart may really be in playing with the toys, but they sold their soul for a few dollars. Some companies provide a "dual career path" strategy that allows the very best engineers to continue to make more money at their first love, but others are trapped in the view that engineers are a simple commodity and that only managers are worth the higher salary.

One consideration sometimes overlooked when an engineer moves to management is that middle managers are legendary for being replaceable. However, a degreed and capable engineer with some amount of personality can always get a job. Somebody always needs design work done or perhaps a sanity check on work done by foreign contractors.

Understanding the Necessary Level

After a few years on the job, engineers may have gained experience in technology, crafting documentation, writing proposals, and perhaps creating and maintaining schedules and budgets. However, to advance their career they must also become adept at social interaction, corporate politics, and building relationships with more senior folks. Great engineers all have a special psychological component. They believe they can and will solve any problem. This internal conviction and drive sometimes comes with the view of life that nobody's opinion is superior to theirs. This self-confidence, critical to solving the most complex problems and finding the most difficult bugs, may also view the opinions and directives of an executive as equivalent to those of a peer or subordinate. This view of life gravely threatens career advancement for engineers. The engineer may not care about the level of the person making a suggestion, but the senior folks are acutely aware of relative levels and chains of command. There are exceptions, but in general executives don't discuss an issue with a subordinate; they give direction and expect it to be followed. If they want your opinion, they will ask for it.

Relative levels are especially important when interacting with other companies. Unless special arrangements have been made, you will be most effective working with your peers in another company. The correct way to escalate an issue is for you to bring it to your boss's attention and have your boss talk to his or her peer in the other company. In many circumstances, a senior executive in another company will not return calls if the caller is not of the appropriate level. Indeed, well-trained secretaries screen the calls of executives to ensure they are not bothered by those of an inappropriate level. Understanding the importance of relative levels and where you fall in the hierarchy is critical to successfully navigating corporate bureaucracy and politics.

In a corporation, it is generally assumed that a senior engineer is better at engineering than a junior engineer. Indeed, it is relatively common for a team leader or manager to assume he has greater skills than those reporting to him. I believe the basis of this assumption is the need for some way for the uninitiated to easily distinguish grades of engineers and management. Engineering, for those not expert in the field, can be indistinguishable from magic. Nevertheless, at times the business leaders of the company must assign individual values to a group of engineers. Seniority is a readily available and visible mechanism to accomplish this. A driving factor behind this need to assign relative value seems to be the desire for a clear chain of command and for staff and management to know who their peers are within and outside the company. Once the hierarchy is established, it becomes unseemly for anyone to act as if a lowly engineer receiving lesser pay has superior skill to a manager.

I'm aware of several cases where an engineer or manager left a company and returned a few years later in a much higher position having successfully bypassed several intervening layers. Accomplishing the same advancement within the company would have required the endorsement of a powerful advocate. In some cases, actual talent has little to do with the advancement. Advancement by leaving the company requires good negotiating skills and a sympathetic ear in the new company. A powerful advocate in your court is a wonderful asset regardless of your talent level. The point here is that having great talent does not guarantee advancement and that having mediocre talent does not preclude it. It should be noted that going to another company does not ensure advancement. Although I know of several cases of great advancement, the vast majority of those who leave their current company have about as much success as those who stay.

Sometimes a gap occurs in the needed hierarchical structure. There are situations when a person is needed to represent a group or to interface with an outside company. Occasionally, such a need is filled by the promotion of a less than stellar individual. Once done, most companies will not take corrective action, since doing so requires paperwork, perhaps confrontation, and admitting a mistake. These accidental leaders continue to cause damage for years because they refuse to learn from subordinates and are not taught by superiors who may tend to ignore the problem. As you walk around the hall and hear things like, "How the hell did *he* get to be a director?" now you will know.

When a customer is upset or when trying to win new business it is important to show commitment to that customer or prospective customer. A traditional approach to this is to have senior-level people attend meetings or visit the customer site. At times, this is not so much about needing their senior expertise as demonstrating that the customer is important to you by devoting expensive people to attend their needs. Important customers don't want to talk to minions, and presenting them with insignificant people would show lack of respect for their value.

All of this corporate focus on the appropriate or necessary level leads inexorably to ever more attention being placed on the title and level of the person you are talking to and the person doing the work. This can be worse in the marketing organization than in engineering. At one company where I worked we used to joke that the entry-level position in marketing was a director. In every group, however, the pressure for position, respect, and titles is relentless. If unchecked, this pressure inexorably leads to title inflation. In large companies it also leads to a huge number of personnel grades. There can be junior managers, senior managers, assistant directors, associate directors, managing directors, and a plethora of other titles. If you are hired into a company as a manager expecting to be one promotion away from being a director, you may be rudely surprised to find that you actually need several promotions.

Advocacy

The corporate chain of command is very important to those near the top of the hierarchy. It should, therefore, be important to you. Offending the senior people is not career enhancing. An important consequence of the typical corporate structure is that a junior or mid-level engineer simply does not bypass layers of management and walk into the office of a senior person and start offering suggestions. Lower-level engineers with a good idea normally need an advocate no matter how good the idea. In general, the rank-and-file working engineer needs an advocate for many phases of his or her career. This advocate is someone present in high-level meetings who can represent the interests of the lower-level engineer. The advocate is someone who speaks up for the engineer when it is time for a promotion or to decide which engineers get a bigger-than-average raise. The advocate does this because the engineer makes his or her life easier or makes him or her look good by helping projects be successful (or perhaps because the engineer is a son-in-law). Obviously, managers or senior engineers will not be the advocate of someone who embarrassed them or caused their project to fail.

For this reason, some caution is advisable when assigned to a difficult or high-risk project. Sometimes project disasters taint all involved even when this is undeserved. Association with a doomed project is not good, and warning people of the impending disaster can be even worse. Sometimes senior people are surprisingly blind to difficulties involved in completing needed tasks. Sometimes this is "purposeful" blindness in the sense that they are wishing very hard for a miracle that allows success. There are also times when they have no clue and in general will not appreciate your trying to teach them.

You need an advocate to assure career advancement. Generally speaking, people will be your advocates only if you make them look good or otherwise improve their life. This fact of life leads to a sequence of incestuous back-scratching from the top of the company to the bottom. This might sound like a distasteful situation, but it only makes sense. Few people will endorse those who belittle or embarrass them. There is little likelihood of this changing as long as human nature is involved. You can choose not to participate, but that will change nothing other than your chance for advancement.

Empowerment and Authorization

For several years workplace *empowerment* has been something of a buzzword. While empowerment is frequently discussed and recommended, such discussions rarely present a corporate power distribution schematic. How does a company determine who gets power and how much they get? When one gets empowered, from whence does the power originate?

The most straightforward representation of corporate power dissemination is the company organization (org) chart. In the simplest sense, if somebody is your boss he or she gets to tell you what to do. If somebody is the boss of your boss, he or she, too, gets to tell you what to do. While various levels may be depicted in the org chart, some of the tiers may also represent significant divisions of power and responsibility. Although important, such consolidation points might not be visible on the org chart itself. This information might be found in the company's personnel policy handbook or even in a more obscure document.

One significant power grouping is in the ability to approve purchase orders. In a larger company, for example, a manager may be able to approve a purchase order of up to $500 whereas a director can approve $25,000 and a vice president can approve $100,000. In a smaller company, every purchase order may need to be approved by the chief financial officer.

Another significant power grouping is the ability to legally bind the corporation to a contract. Such binding power is not restricted to corporate attorneys. It is restricted to folks called *officers of the company*. An officer of the company may have a variety of titles, including but not limited to *vice president, attorney,* and *director*. Under most circumstances, an engineer signing a contract does not legally bind the company to the terms of the contract. However, an engineer signing a bad contract could generate a legal headache for the company as he or she tries to maintain good will and dodge accusations of setting false expectations.

When various books discuss empowering an engineer, assembly line worker, secretary, or project manager they are not talking about empowering employees to sign purchase orders or contracts. They are talking about empowering employees to take charge of their lives and to make decisions about their personal work. Employees may also be authorized to make decisions as a manager's representative or substitute. Authorization is a little different from empowerment. A person is authorized to do things and make decisions that affect other people such as signing a purchase order or deciding the color of a new product. A fairly common problem is that impromptu authorization given to an employee may be unknown to the remainder of the organization. I've been in many meetings where a senior person authorizes an engineer to act as her representative or take some action but neglects to tell others about the authorization. The engineer soon finds himself in the difficult position of giving direction or recommendations that are not respected. Due to political infighting, there are also situations where proper notice of the authorization is provided but is simply ignored by those with a different agenda.

Telling me I am in charge of something is not the same thing as telling everyone else I'm in charge. In fact, telling everyone I'm in charge is not the same as everyone knowing, understanding, and accepting that I'm in charge. There are occasional bumps in the dissemination of impromptu and delegated authorizations, even in companies with well-established and mature processes.

Caesar and the Engineer

Shockingly, engineers may discover that not all of their colleagues are nice people. A small number of coworkers and managers view projects and engineers as commodities to be conquered and controlled. In extreme cases, these *petite Caesars* create a sphere of reality-warping influence that disrupts productivity and innovation. In such an environment, truth, engineering, and quality can be irrelevant. Intellectually gifted people become feared instead of treasured. Innovation can be viewed with skepticism, and risk must be avoided at all costs. Dissenting opinions are ruthlessly purged. At the most fundamental level the priority becomes supporting the Caesar and making him—or her—look good. Statements made in ignorance or confusion become *de facto* rules. When this type of environment spirals out of control, designs are bent to his or her view of reality. The spotlight shines only on Caesar, who gets the credit and the glory, but never the blame.

Not many situations get as bad as that just described, but there are "wannabe" Caesars everywhere. One of the most commonly observed phenomena associated with this is *Caesar marking*. Caesar marking is similar to the marking that may be done by a dog or cat. It shows possession and ownership of turf. The marking can be something as simple as a few meaningless edits to your document or as egregious as a movie director's changing the decades-old storyline of a comic book on which the movie is based. Is any real value added to changing the paragraph indentation or a few words in a document? Were the movie script writers really so inept that they couldn't write a good story without changing the venerable origins of popular comic book characters? Why is this done? First, because they can. Second, it shows their dominance and control. Third, it cultivates the image of adding value.

Everything in a Caesar's view must have his or her mark. This incessant need to demonstrate dominance (or perhaps value) slows productivity and creates indecision in the staff. The employees' continual fear of being second-guessed paralyzes creativity and in extreme cases can even damage the work ethic. Why should someone work hard when his or her work will be criticized and redone anyway?

Managers and Motivational Techniques

Being a manager of technical people can be difficult. One may often hear it compared to the job of herding cats. It can also resemble the job of parenting children. Perhaps an idealized view of the job of a technical manager is providing parental oversight and mentoring to gifted children. Good management encourages the staff to think, grow, and excel. Unfortunately, the real world can fall a little short of this view. Sometimes the problem is that of the manager, and sometimes it is the result of events and the environment. It is important to remember it is the job of managers to do everything they can to keep projects on schedule and to hold costs down.

Management simultaneously holds several somewhat contradictory views of engineers. One is that they will work on a project forever if you don't pry it from their hands. This is the "perfect airplane never leaves the hanger" paradigm. Another is that the engineers always want to build complicated geeky features into the project. Another is that engineers never willingly incorporate desirable but hard-to-implement features into a project. On any given day, in any given argument, the stereotype that surfaces may be the one that best supports the wishes of the manager in charge.

Some technical managers view their biggest challenge as motivating the staff to work hard to implement the specified features in a timely fashion. At times there can be a management perception that engineers are just not committed to delivering the features on the stated schedule. Sometimes the pragmatic and negative consequences of certain motivational techniques are not obvious or not considered.

There are as many motivational techniques as there are managers. Some techniques are taught in schools and seminars, and others are learned by experience. Others simply exist despite extensive attempts to banish them. A seemingly endless array of questionable techniques is used by desperate managers in a generally futile attempt to keep engineers on hopeless schedules.

Cheerleading

Perhaps the most innocuous of the problematic motivational techniques, cheerleading may also be the greatest double-edged sword in the manager's arsenal. Managers and executives are supposed to excite and motivate the corporate employees to perform to the limit of their abilities. This not only makes for productive employees but also keeps morale high. There is, however, a fine line between optimism and insanity. Some executives are habitual cheerleaders. They will never say a serious problem has been encountered or consider giving up. In some cases, they will continue to say positive and encouraging things even while they desperately work behind the scenes to create contingency plans. In some cases, the executive really is blissfully unaware of the severity of the problems being faced. In either case, many of those on the front lines who are in direct contact with the real world cannot be fooled.

Managers lose credibility when they remain hopelessly optimistic in the face of real-world facts. They may be able to continue to manipulate the hopes and goals of the clueless and naive, but they will lose the respect of those capable of thinking for themselves. There seems to be a natural contention between boisterous managers who attempt to lead with bravado and the intellectually gifted thinkers. Bravado may work well in politics and with a poorly educated population, but solving hard engineering problems requires the most gifted our society has to offer. Irrational cheerleading risks losing the respect of the best and brightest. Cheerleading intended to inspire and motivate may condemn to failure a wounded but salvageable project.

Management by Small Progress

From time to time various members of the engineering staff may become negative or express negative viewpoints about a technology or schedule. Management fears that such negative viewpoints might become infectious and cause people to lose incentive. They may attempt to combat this problem by forcing the staff to ignore risks at the end of the project and to focus on successfully completing short-term goals. The project can then be effectively advanced by small steps. Superficially, this small progress approach makes a great deal of sense. Putting distractions out of your head and working on the job at hand really helps to get that job done. From a higher perspective, management by small progress makes no sense whatsoever. If the concerns about the project are real, ignoring them and working on short-term goals is like rearranging the deck chairs on the Titanic. You're sunk, but you will deliver a few more milestones before you drown.

If the concerns of the engineers are ultimately justified it makes little sense to ignore them and focus on short-term problems. The difficulty for management is determining whether the engineering concerns are real. This is an especially difficult problem if managers have no personal expertise in the problem areas. I know of a case where a senior engineer had very serious concerns about the viability of a project plan. He went to the head of the engineering division and presented the case for what was feasible, what was unlikely, and what was impossible. In effect, he explained in detail why the current plan would lead to failure. The division head, who listened but was unconvinced, said wistfully, "If only these problems were really so obvious." Said the engineer, "Well, they are to me."

The project actually did fail as predicted by the engineer. Management lacked the subject expertise and years of experience to see what was clearly visible to the engineer. Worse, although they personally knew little about the technology they did not trust the educated viewpoint of their own expert. They continued to focus on short-term goals and small progress until the project sank. What was most unfortunate was that a great many of the short-term achievements had to be thrown away when the project was redone with a different approach. The money and time spent on the first project was nearly a complete waste.

Am I saying that management by small progress is a bad idea? No. What I am saying is that intelligent and skilled people usually have worries and concerns for good reason. Just because the manager doesn't understand them doesn't mean the concern isn't warranted. In a fair world engineers and managers who foresee distant problems would be honored and rewarded. Unfortunately, such visionaries can be considered a negative influence, and those who suffer the technological equivalent of prosopagnosia, or face blindness, are often considered motivational leaders.

Management by Focus

Management by focus is similar to management by small progress in that it restricts the view of the project to near-term goals. Management by focus is different from management by small progress in that the latter usually comes about when some members of the team begin expressing concern about the project. Management by small progress is an intervention intended to salvage a project, whereas management by focus is the primary methodology of the project. The project is purposefully initiated with an impaired view of the long-term goals. This approach is generally used when management has not yet defined much of the project and doesn't want the engineers to worry about this. The management position is that the engineers will be told about the next phase when they need to know. The reality is often that management needs to figure out the next phase.

Engineers with technical vision can react badly to a project "focused" in this fashion. They worry that designs done with incomplete knowledge of the overall system may be inefficient or even incorrect. Their concern may be met with some hostility by the focusing manager. Questions by the engineer can expose the lack of a plan or challenge the dominance and control of the manager. In such situations it is good for engineers to remember that they are paid to help bosses be successful and surprising or embarrassing them is not career enhancing.

Management by Ambiguity

For the engineer, management by ambiguity is perhaps the most irritating form of well intentioned management. It can be used when the person in charge wants to enhance the creativity of his staff. He attempts to spur out-of-the-box thinking by removing the sides of the box. He does this by providing a large number of "requirements," some of which may be more properly termed "goals," some of which are nearly impossible, and some of which are truly essential. He then demands that the staff produce a design but allows them the freedom to disregard requirements of their choosing to meet the cost or schedule allocated to the project.

The manager forces creativity by steadfastly refusing to identify which "requirements" can be disregarded and which are absolute necessities. He instructs the staff to give him a design and tell him which "requirements" will not be met. An endless loop of failed design reviews ensues when the manager rejects design after design that does not accommodate the features he knows he needs. You may wonder how any manager could do something so silly but I have seen many smart managers get caught in this trap. It indicates a lack of faith in the ability of the staff and inadvertently creates an environment where success is not determined by the engineer's technical skill but by his or her psychic ability to read the manager's mind and discern the real needs.

Unfortunately, this form of management often becomes a self-fulfilling prophecy, and pushes the team into creative bankruptcy. The manager started with the suspicion or belief that members of his staff lacked creativity, but then crippled engineering innovation and progress by allowing too many degrees of freedom. It can be the really creative engineers who are most damaged by this management strategy. They see so many options and possibilities that they need to know the REAL requirements to contain the diversity of design options. They flounder in a sea of infinite choices, needing constraints to anchor them.

A cousin of this approach is to provide minimal requirements so as to not restrict creativity in any way. The results are the same. The engineers and manager are caught in a dance of frustration where each step further convinces the manager that his people just don't know how to design and are not very imaginative. The manager believes he did everything possible to stimulate innovation but neither an avalanche of requirements nor a dearth of requirements furthers this goal.

Given enough time and money, nearly anything can be engineered. The real world of engineering design is a myriad of tradeoffs between cost, schedule, features, quality, polish, and professionalism. In the real world, engineers must know the genuine minimal requirements to evaluate design tradeoffs. Managers who refuse to identify critical requirements do so hoping to encourage the staff to reach for the brass ring and achieve greatness. Unfortunately, they more often achieve the opposite.

Management by Secrecy

Knowledge is power. Those who control access to information control their enemies. Management by secrecy is a technique used to maintain control over people and projects. Withholding information prevents people from intelligently challenging decisions and direction. Withhold enough information, and people don't even know a decision was made or there was something to challenge. Usage of this technique is extremely widespread and includes many variations:

- Hiding component pricing or specifications that support an out-of-favor design
- Not telling certain people about a meeting so that those who do attend the meeting can make the ordained decision
- Scheduling a meeting at a time or location that makes it difficult for rivals to attend
- "Forgetting" to send an important memo to the adversaries
- Suppressing customer complaints or product failures that make a favored organization look bad
- Declining to respond to questions about an issue
- Selectively releasing only that information and evidence that supports the favored viewpoint

- Merging a successful division with a money-losing division to hide the losses
- Appointing a new leader of a disparaged group while allowing or perhaps encouraging the old offensive priorities, culture, or practices to continue

There are many more incarnations of this technique. Zen masters of management by secrecy skillfully avoid challenges and easily get their agenda, personnel, and design adopted. Clever use of management by secrecy avoids the need to actually lie. The victim is merely misdirected to the intended belief. In some cases, continual repetition of the same half-truth convinces many people that it is accepted fact. In other cases, management can lie to some extent because the information needed to expose the lie is hidden. This type of behavior can and does thrive in corporations because there is no sentinel tasked with detecting such behavior. There is no "free press" to keep the managers honest.

Unfortunately, secrecy and accountability make poor bedfellows. Secrecy confounds accountability by preventing debate of the hidden issues. There is no peer review, no fresh ideas, no opportunity for improvement, and little likelihood of punishing wrongdoers. Depending on self-policing or self-regulation seems very high risk. We've seen that even the Catholic Church and congressional leadership may occasionally look the other way with issues as severe as molestation of children. Where may one hope to find leaders more righteous than those charged with shepherding of souls or leading the free world? Practical experience has repeatedly proven that self-policing cannot be trusted. J. Robert Oppenheimer said it well:

> We do not believe any group of men adequate enough or wise enough to operate without scrutiny or without criticism. We know that the only way to avoid error is to detect it, that the only way to detect it is to be free to inquire. We know that in secrecy error undetected will flourish and subvert.

Secretive and deceitful behaviors are used, frankly, because they work and work well. Most people don't expect smokescreens, insincerity, and misleading practices from those in leadership positions. This naiveté allows ready acceptance of the agenda being promoted. Aggressive intelligent people can observe the success of these deceptive practices and adopt them for their own use. In some workplaces, this results in an ever increasing spiral of secrecy, deception, and hidden agendas. In these environments, discovery of an impropriety is usually met with denial. Denial may continue until an extraordinary amount of evidence has been amassed and presented. Under great pressure, denial may turn to apology, but such apology can often be insincere. The offenders are sorry only that they were caught.

Under no circumstances am I suggesting that deceptive practices are in widespread use by executives. What I am saying is that employees should not be surprised to encounter this type of activity. It exists and occurs to some extent in many corporations. As a working engineer, however, it may not be career enhancing for you to expose uncovered deceptive practices. Actively pursuing such issues and aggressively attempting to force account-ability may ruffle important feathers. You could be identified as a trouble-maker or naysayer, and your activities could be viewed as a witch hunt. It is difficult to demand accountability when a superior states, "We don't appre-ciate finger-pointing in this organization." Accountability can be mandated only from above. An environment in which executives readily admit that "mistakes were made" but in which no consequences result is an environ-ment that creates a powerful deterrent to successfully holding individuals responsible for inaccuracies or unethical behavior.

Management by Misdirection

Magicians and some managers make heavy use of misdirection. Misdi-rection allows the user to seemingly accomplish magical feats while using ordinary means. The ordinary means are not noticed because the attention of the observer is purposely directed elsewhere. Management by misdirec-tion is a close relative of management by secrecy. The two methods share a large number of characteristics, and some of the techniques listed previously under management by secrecy may be argued to more accurately be manage-ment by misdirection. I make the distinction that management by misdirec-tion provides a specific distraction whereas management by secrecy merely omits or hides information; that is, in the former, the victim is specifically led to believe something inaccurate whereas the latter provides no information at all and allows the victim to be blissfully ignorant and unaware.

Examples of misdirection are legion. Consider these real-world examples:

- A project status report states that the hardware group is waiting for a component to arrive from another company. Though this may be true, the report neglects to indicate that hardware development is not yet ready to use the component. An irrelevant event is implied to justify delays experienced during product development. This status report also made no mention that the hardware delay is adversely impacting software development. Score a misdirection double vic-tory for the hardware manager. Not only does he get a free pass for being late with hardware development, but his responsibility for making software late has also been obscured.
- A test report states that a product under trial successfully passed 27 tests. It neglects to mention that it also failed several tests, one of them catastrophically. The catastrophic failure damaged several thousand dollars worth of test equipment.
- A corporation was negotiating the lease of a new office building. Employees working at the old building knew for many months

that they would be moving. A number of them periodically asked the location of the new office. The answers they received varied greatly. Some of the answers alarmed staff members, who thought they would have to drive much farther to their job. It turns out that misleading answers were purposely given to landlords to convey the false impression that a decision was made. This misdirection was used as leverage to negotiate a better price on the desired office building nearby. To maintain a consistent story, management had to give the same misinformation to the employees and to the landlords of the desired office building.

- A status report states that the new application-specific integrated circuit (ASIC) arrived on schedule. It omitted the fact that it didn't work.

- Congratulations are shared when the news is announced that a new product shipped on schedule. It shipped to the brother of the director of engineering, and he was warned to not turn it on for fear of fire.

- It is reported that the focus group loved the new user interface. No mention was made that the PC running the user interface demo froze up every few minutes and had to be repeatedly rebooted.

Management by misdirection is frequently used at nearly every company. Skilled practitioners use it to avoid responsibility for problems and to share in the glow of success. Penetrating the deception may require intuition and sufficient subject knowledge to ask meaningful questions. Persistence may be required to keep asking questions until the real truth surfaces and the misdirection of the magician is exposed. Unfortunately, exposing the misdirection may not be career enhancing because the exposed manager may bear a grudge and others may take offense at the aggressive tactics used in detecting it.

Management by Pressure

Management by pressure may very well be the most common form of supervisory technique. Managers and executives can be very busy. They don't always have time to understand project details or to appreciate technical nuances that are causing delay. It is much more expedient to simply insist something must be done than to actually understand why it can't be. Furthermore, managers who demand something be done demonstrate they are in charge and are properly motivating their personnel. It is unacceptable for managers to stand by and allow a project to encounter delay after delay.

There are only three developmental reasons that a project might fall behind schedule: (1) the problems being worked are harder than anticipated; (2) the assigned engineers are not very good or some tasks are out of their area of expertise; or (3) the engineers are not working hard enough on the project either because they are lazy or because something is conflicting with this effort. All other reasons for delay are some combination of these reasons or delays in receiving components from a vendor. The good news for managers is that applying pressure addresses all of these possibilities. A lazy engineer will be motivated, and the project of the manager who applies the most pressure will get priority. Sufficient pressure will also force the engineer to learn faster and work harder to solve the problems. Pressure is seemingly a universal solution to schedule delays.

When a serious problem is encountered and progress halts, the manager could ask the engineer when it will be solved. Unfortunately, experience teaches the manager that the answer is often the most feared and unacceptable "I don't know." It is much easier to just tell the engineer the problem has to be fixed by a certain date. The engineer must then be managed to help him succeed in this task. The most obvious way to do this is to periodically remind the engineer of the importance of achieving the stated date. This could mean having the engineer do hourly progress reports. The manager may also hold daily meetings to meticulously review all issues and to verify the engineer is working on the correct things. The engineer simply cannot be allowed to miss deadline after deadline. If the deadline is missed, the biggest weapon left in the manager's arsenal is even greater pressure.

It is worth noting that there are a few negatives to severe pressure. One possibility is that the engineer may tire of the stress and find another job. Another possibility is that the engineer may become less concerned about doing things right and more concerned about doing things quickly. Stupid engineering tricks are all about doing things too quickly. Most people are pain averse, and engineers are no different. After weeks and months of pressure anyone may become quite cavalier in decreeing a problem solved. For engineers who remain conscientious and true to their profession, the only option is to work longer hours.

I have been managed by pressure on numerous occasions—so often, in fact, that I believe it has allowed me a fresh perspective on the need for sleep. Neuroscientist and philosophers continually look for the benefits of sleep. They theorize that there must be an incredible benefit from sleep because of the serious disadvantages its lack presents. A sleeping animal can't eat or reproduce and is more vulnerable to predators. There is even an indication that mammals deprived of sleep die faster than their well-rested peers. This is persuasive evidence that there must be a very good reason for sleep. I agree, but I think the researchers have been looking at this the wrong way. The advantage of sleep is not in resting and repairing the body or in memory and cognitive remapping. These, in my opinion, are opportunistic functions taking advantage of the existence of sleep and are not the root cause of sleep. The real reason we have sleep is to provide a reservoir of extra time. There is designed-in pain associated with tapping into this reservoir of time. This pain is proportional to how much "extra" time we need.

I can say with confidence that the existence of sleep is a savior. That is, if not for the ability to temporarily reduce the time we sleep we might very well perish when times get difficult. Where would we find extra needed time if our day were already filled to capacity with activities? Perhaps a few of us would be smart enough to allow some slack time in our lives, but some managers I have worked for would only push harder. In times of famine extra hours are needed to scavenge for food. When the enemy is at hand the Roman soldiers must be on the march. When the project is late we must work later. Sleep, my friend, exists to fill up free time and to provide a time reserve to bail us out when we overcommit.

Management by Coercion

Sometimes management by pressure becomes extreme. It could involve threatening your employment, bullying, or even physical threats to your person. Methods of coercion are as limitless as the imagination of the executive in charge. Management by coercion steps over the line from pressure to threatening behavior. It is interesting to observe that management by coercion can be a powerful incentive to work hard, and it can also be effective at preventing people from visibly deviating from the approved party line. Coercion can readily influence many people to obey instructions and to avoid disruptive behavior.

Managing Up

Managing up is something of a misnomer. The term is generally used to describe the process of managing your boss, but my experience is that such so-called management can more closely resemble following direction and kowtowing to the party line. Some supervisors maintain a functional rapport with their staff. Little managing up is needed in these situations because management and staff are on the same page (or at least the same chapter). Other supervisors have a different view of life. They give orders and expect them to be followed. They tell you when a project is supposed to be done and don't want to hear that it's impossible. These supervisors are the ones most employees believe need to be managed, but such management may only consist of saying "yes," "okay," and "no problem." Defying or challenging your boss risks raising his or her ire and may result in a contentious argument. When engineers argue with their bosses, it is the engineers who lose.

Timid and insecure people know instinctively that arguing with a superior is a bad idea. Self-assured engineers may be very comfortable telling their bosses their ideas are ludicrous. Upwardly mobile engineers, however, understands that a manager's ego may be bruised when a subordinate rejects an idea as inferior. Explaining why it is inferior just makes things worse. The world can be a complicated place, but your manager may prefer to think of it as simple. Experience teaches that effectively managing up involves supporting your supervisor's view of the world. It is best to roughly approximate his or her view of the complexity of issues and the severity of problems. Sometimes when your boss's suggestion is the most stupid thing you ever heard, the right answer is, "Good idea, thanks."

Agreeing with bosses and appearing to share their worldview buys you credibility and latitude. Managers tend to focus their time and effort on people and projects they perceive to be problems. This means that projects that are behind schedule and individuals who don't follow orders get more attention. More attention means more status meetings and more help and counseling. Often this amounts to less engineering time being devoted to actually solving the problems because time is consumed preparing for and sitting in meetings. Although counterintuitive to many managers, sometimes the fastest way to get a problem solved is to stay out of the way and let the engineers work on it. Managers are more likely to do this if they trust the engineers—and they are more likely to trust those who seem to have a shared belief system.

Patterns and Portents

The majority of people in most companies are not engineers and lack sufficient technical background to understand if a design should take 2 days or 2 months. To them, the engineering technobabble is indistinguishable from a magician's spells or a politician's mumbo jumbo. All this makes it difficult to believe engineers when they say something is going to be a problem. To the average person it is impossible to differentiate engineers from soothsayers looking into a crystal ball. It all seems so magical that there is really no reason to believe engineers. Experienced technical people may see a clear pattern that is not discernible to the layperson. This pattern can be a portent of looming problems but may be overlooked by those not sensitive to the mystical energies emitted by sophisticated technological devices.

Ideas and Designs

Experienced engineers know there is a big difference between having an idea and having a complete design. Many times, technically inexperienced people or people who only manage or dabble in technology vastly underestimate the difficulty of progressing from an idea of how to do something to a robust, reliable, and reproducible design. Robust, reliable, and reproducible are the three magic Rs of engineering. Being able to build something that works dependably with production variation of components over real-world environmental conditions in an environment of oblivious users challenges and at times befuddles the best minds of the human race. It is far, far more difficult to build 1,000 copies of something that works for 1,000 users than it is to build one thing that works for one user. In the business world, far, far more difficult directly equates to much more expensive.

People who don't think like an engineer sometimes confuse believing something and knowing something. You can believe many things to be true. You can believe that hot water in ice cube trays freezes faster than cold (why else would the Zamboni machine spread hot water on the ice rink?), or you could believe your thermodynamics instructor who says that cold water freezes faster. The real engineer says belief is irrelevant and tests the hypothesis. Modern engineering makes use of some sophisticated and (generally) trustworthy equipment to run simulations instead of actually running a test. The important thing is that real engineers don't accept that something is working just because it seems to be working. They also know that seeing one of your products work at –40°F doesn't mean the next 10 will. Quality design, solid implementation, and thorough testing separate confidently knowing something works from merely believing it works.

Unfortunately, the workplace can experience significant confusion between believing and knowing and ideas and designs. This difference in thinking can be a function of one's basic personality or the result of a disparity in depth of understanding. Sometimes the split is along cultural lines and result in engineers, managers, and financial folks' having strikingly divergent views of a situation. Bridging this gap and negotiating compromises can be some of the biggest challenges for the working engineer. At times it can feel like explaining the science and evidence that proves the world is round to zealous adherents to a religion that teaches the world is flat. Presenting facts results only in the zealots' shutting their eyes more tightly and believing more ardently whatever they want. Most infuriating is that no matter how many times one of their beliefs is proven wrong they cling just as passionately to the next one.

Prototypes, Demonstrations, and Products

As there is a big difference between having an idea and completing a design, there is a big difference between doing a demo and shipping a good product. Demos can be, and often are, barely working and held together with paper clips and duct tape. Sometimes the mad rush to do a demo severely damages the structure and quality of the actual product. Likewise, a prototype exists because there were questions and uncertainty that needed to be verified. If engineers already had all the answers they would not need to make a prototype. Not everyone in the company understands the amount of work, and therefore cost, remaining to convert a prototype or demo into a final product. This lack of understanding can result in unreasonable expectations and contention. As the pressure on the engineering staff members builds, they may resort to workarounds and patches instead of tracking down and fixing the root cause of problems.

Doing any development—a prototype, demo, or product—is far harder, more expensive, and more time consuming in a business than it is in a capable engineer's basement. To those not working for a company as an engineer this must seem extremely counterintuitive. It is nonetheless real. In a company, a great deal of effort and expense must be devoted to getting approvals, assigning engineering staff to a project, getting charge numbers, building consensus, and 100 other bureaucratic tasks. An absolutely staggering amount of additional work is needed to develop a product in a large corporation when compared with doing the same work in your basement. This discrepancy between the total amount of work and the work actually devoted to the project can be large and invisible to those not intimate with the activities. This overhead can be completely overlooked, causing important people to become impatient and mystified as to why the engineering takes so long.

Corporate executives can have a great deal of confidence that they understand the true status of a project, but few have the necessary background and are close enough to the problems for this confidence to be justified. I was once horrified to hear the president of my company telling investors my project would be completed in a few weeks. Later, in private, I asked why he believed I would be done so soon. He assured me that he had been carefully tracking my project and had an accurate view of it. I couldn't even imagine how he convinced himself of this. He had not spoken with me or my team. In reality, he had no idea how much work remained to complete the project.

Other Options

Engineers do many more things on the job than "engineering work." They can make coffee in the office kitchen or call maintenance when the urinal overflows. We've discussed a number of bureaucratic tasks, but a variety of other activities not directly associated with designing or building a product also make use of engineering knowledge.

Publications, Presentations, and Patents

A great deal of prestige can be associated with presenting at a conference, having a magazine publish your article, or being awarded a patent. All these things look good on your resume and provide exposure of your name and skills. However, as a condition of employment, many companies require an engineer to sign away all rights to inventions and to notify the company immediately of new ideas by filing an *invention disclosure*. In effect, the company owns its engineers' brains, and it is the company that files the patent, not the engineers. Likewise, writing a magazine article or doing a seminar presentation about your work most likely requires approval of your company. Such approval can be delayed or denied for several reasons:

- The company may wish to keep the work secret. They may not want others to know much about products or technology being developed.

- Those who must approve the article or presentation may be too busy to review it. While people are usually busy, not being able to find time to review an article for publication or a patent for submission to the patent office may have an ulterior motive. As engineers improve their resume and get more exposure it becomes harder to deny them promotions and raises. It is not in the corporate management's interest to allow this to happen. Certain cultures and individuals may try to keep engineers nameless, faceless, and obscure.

- Individuals may feel threatened. Some might enjoy a position in the spotlight as the one who writes articles, files patents, or represents the company at conferences and might not appreciate others sharing in the spotlight.

Patents are a little special in that they represent intellectual property, and some are extremely lucrative. A number of companies make a great deal of money from the licensing of intellectual property. Unfortunately, it can be difficult to know in advance which patents will end up being important and financially rewarding. I know of no company that offers engineers a portion of profit derived from patent royalties, but many companies offer token amounts of money to encourage their engineers to write patents.

I'm aware of one case in which an engineer dutifully filed an invention disclosure notifying his company of his possibly patentable idea. The company decided that the idea was without merit and declined to pursue filing a patent. However, the engineer believed the idea was sound and potentially valuable and decided to personally file a patent. He fulfilled his obligation to notify the company about the idea and believed that he was free to pursue the patent once the company decided not to. The company saw things otherwise. They threatened the engineer with legal action. The company's position was that the engineer had signed away all rights to the idea and it was solely the company's choice to either patent or ignore it.

There can also be political intrigue with patents. Some managers encourage collaboration and teamwork and instruct engineers to share the honor of a patent by adding the names of coworkers and supervisors to it. I am familiar with one less than honorable manager who instructed a creative engineer in another group to add names of key members of his group to a patent. The manager explained that it was the corporate culture to share patents in this fashion. Several months later the engineer was very surprised to learn that one of the people gratuitously added to his patent neglected to return the favor. He discovered that the dishonorable manager advanced the cause of his group by insisting that others include them in honors but did not extend the same courtesy. The guise of teamwork can be used by immoral managers eager to share in the success and spotlight of others but unwilling to share theirs.

Bids and Proposals

Some engineering jobs involve contributing to a large number of proposals, and some do not. It depends on the business of your company and the nature of your work. Marketing and sales departments work hard to find prospective customers, but engineers may get involved in determining how long the proposed development will take and how much it will cost. For the context of this book, the more relevant situation is when these engineering estimates are minimized or ignored.

In a number of corporate cultures it is a generally accepted fact that all companies bidding on a development contract are exaggerating their capabilities to win the work. Therefore, any company hoping to be competitive must do the same. The theory is that once the customer has chosen a development partner it is unlikely to go through the pain, effort, and cost to change partners. A corollary to this theory is that all you have to do is not be too late or aggravate the customer too much and you can keep the business. How does one keep from being disastrously late on a purposefully optimistic schedule? Easy: Beat the engineers to work harder and faster and blame as many delays as possible on indecision and requirement changes by the customer. How do you keep from losing money when you used an excessively low price estimate to win the business? Easy: Dream up reasons to charge the customer more money for added features, design modifications, and everything else you can think of.

Interviewing Job Candidates

I'd learned some years ago that many (most?) employees tend to agree with the senior person's assessment of a job candidate regardless of their personal feelings. Knowing this, I never reveal my opinion until the more junior people express their view of a candidate. On one interesting occasion I and four engineers in my group interviewed a prospective addition to the team. The candidate badly failed my interview by giving nonsensical responses to a few technical questions. I was amazed when all four of my engineers said they liked the candidate. I gently probed with some obtuse questions. Do you think he could do this work? Do you think he could do that work? It turns out that all four merely asked the candidate to talk about his experience. He told them all the great things he had done.

At this point I realized that the team needed some training in interviewing job candidates. It seemed that nobody had actually asked technical questions. None of my engineers had bothered to challenge the candidate to prove the knowledge he claimed. It is difficult to determine the authenticity of someone's knowledge if you don't have personal expertise in that field, but if you do it is easy. I once asked a person detailed questions for over an hour and got only vague answers or sweeping generalities about the work. In frustration, I finally said, "Look, sooner or later I have to understand whether you know this stuff or you sat next to the person who knows this stuff." After a few more minutes he admitted that other people on the "team" had actually done the more significant design and implementation. He presented himself well but was only a poser.

It's bad when candidates can't answer job-related questions, but for me it goes downhill quickly when they make up answers and state them with great confidence. It seems unfortunate that some people have learned that they can say bizarre things and have people believe them if they are said with enough confidence. The world created by the people of Earth has few intellectual checks and balances. Civic and moral leaders and "expert" spokespeople can say amazingly absurd things and amass a large following of believers and supporters. In this world, knowledgeable people who challenge the absurdity can be shouted down or shot down. History has repeatedly shown that assertive fanatics can gain control of entire societies, sometimes for centuries. Engineers are fortunate to live in a much more constrained world where fantasy and pretense can easily be detected. It is easy to ascertain if people know what they are doing technically. People need only to have personal expertise in the field. What is much harder is determining their work ethic, how creative they are, and their interest in continuing to learn new things.

Prospective employers should ask technical questions in an engineering job interview, and it is important to distinguish good questions from minutiae. Candidates who don't know what DVT means (design, validate, test) may still be the perfect employee. It could simply be that different names were used for concepts or that they happened to have not been exposed to environments where knowing certain things were important. One interview approach is to focus on questions that reveal candidates' thinking process, but then interviewers must be adept enough to know whether the exposed thinking is good or bad.

I have also noticed that some hiring managers focus heavily on the number of years of experience with a specific technology. That is, the manager tells human resources he or she needs someone with 5 years of experience programming COBOL. The manager's implicit assumption is that someone with 5 years of COBOL is better than someone with 3 years of COBOL. In my humble opinion, the manager is abdicating his or her responsibility to hire the best person by substituting experience for skill. The fact is that an engineer asymptotically approaches competent knowledge of any given technology after using it for a few months. Independent of experience, the better-thinking engineer generally produces better results. Although overall diversity of experience is extremely beneficial, 2, 3, or 5 years of experience with a specific technology is irrelevant. Some managers fall back on the number of years of experience because it is much more expedient and far easier to defend if the new hire ends up being unsuitable.

Another experience-related phenomenon can be observed in the hiring of managers. Many companies have little interest in hiring a manager who has no previous experience managing people. Perhaps in conflict with this, some executives and corporate cultures only consider hiring managers who are technically competent and personally capable of designing and implementing systems. Once hired, the manager is loaded with personnel and project management responsibilities that preclude hands-on participation in technical development. After a couple of years in such an environment, the manager's technical skills weaken and become out of date. Shortly after, the manager becomes a bureaucrat unable to constructively contribute to technical work. It appears, therefore, that companies destroy in a couple of years one of the characteristics they consider most valuable in hiring new managers. As a corollary, there seems to be only a narrow window when managers are most valuable and most marketable—that is, the small number of years between when someone has given them a chance to manage people and before their skills have aged and degenerated. Corollary number two is that it is in managers' best career interest to stay up to date technically and prolong their golden years of marketability indefinitely.

How much better it would be if only the intention of all those interviewing candidates were to ascertain their skill and compatibility with the culture and goals of the organization. Unfortunately, when humans are involved there is no getting away from political intrigue and personal agendas. I was having a friendly lunch with a senior executive friend. He complained at length that there seemed to be no good job candidates for some openings he had. Everyone he saw was inept or worse. He despaired at the lack of skills and meager intelligence of the local population. While I tend to agree that gifted engineers are rare, I find that they exist and can be readily found. Given this experience I couldn't understand the problem my friend was having. After more discussion I began to comprehend.

Like many organizations, he had in place an interview hierarchy. The staff interviews the candidates first, and only if they all liked the person would the busy executive take the time to talk to the candidate. This type of situation usually shows up as "If Available" for the last person on the interview schedule. Hint: Your interview probably didn't go very well when the "If Available" last person isn't available.

In my friend's case, members of his staff were rejecting candidates that were too good. They were rejecting people who would be competition for the next promotion or who were more skillful and could make them look bad. My friend had never considered this possibility. After our discussion he reviewed the resumes of recently rejected folks and asked the most promising to come back for another interview. He was amazed at the quality of the people who were rejected and was extremely disappointed in the behavior of his staff. Quality engineers existed, but they were being rejected by insecure employees who wanted to minimize the competition.

Marketing Support and Collaboration

It is often a good idea to have a pre-meeting or a dry run to walk through issues and presentation in advance of an important customer meeting. This allows time to correct deficient documentation and demonstrations or to make sure all the players are on the same page. On several occasions, I've seen the manager in charge of the customer relationship pick up the stack of documentation being given to the customer to feel the weight of it. The attending engineers affectionately called this the documentation "weigh-in." Was it too light? Customers have to feel that their money was well spent and that enough work was done.

Managing customers and their expectations is critical. A big part of this is making sure that every contract deliverable is provided to the customer in a professional and attractive fashion. On one occasion the program manager didn't notice that one of our required deliverables was a pseudo-code definition of the system. Fortunately, this oversight was caught a few weeks before the end of the contract. I was paid to write a program that parsed a directory full of program source code and generated a very tall stack of pseudo-code. Delivery met.

It is difficult to get more money from unhappy customers. Delivering all the contractual requirements in a professional fashion is important, but equally important is actually solving their problems. There is a truism in customer relations that says give customers what they need, not what they ask for. While marketing can do a fine job echoing customers' stated needs, it may struggle to see the bigger technical picture. Engineering can greatly assist marketing by helping to visualize how near-future innovations can help the customer in ways not obvious to the technically less sophisticated. However, at the end of the day, engineering must deliver what the contract says. Delivering something different to "help" the customer can result in serious legal problems. The desire to help the customer must be tempered with the legal requirement to meet the specifications of the contract.

Image

Individuals constantly create and present an aura. Specific components of the aura such as confidence or shyness might be identified, but the overall image the person conveys can be very complex and at times conflicted. The image and persona may shift and refract and can be quite amorphous. Nevertheless, every individual creates and leaves an impression with all those encountered. A communion of such impressions constitutes the overall image. This can be subconscious— even visceral—and may be positive or negative. The positive or negative connotation of the image can be dependent on the corporate culture and the predispositions of the observer. An engineer may be seen as an innovative genius at one company and a disruptive maverick at another.

Professionalism

We can all agree that being viewed as a professional is advantageous to one's career, but then we must decide what it means to be professional for a given company and position. It can mean dressing well and being respectful to your superiors. It can mean understanding business etiquette and adhering to those generally unspoken rules. Being professional can mean a variety of things and can vary from job to job. Getting a promotion often does not depend on being smarter or working harder than your peers but on creating the impression that you are suited for the next level.

I once worked for a very demanding boss who had an excellent memory. Sometimes meetings were very stressful as the boss would establish dominance and embarrass folks by asking off-the-wall questions about project minutia. A coworker developed a remarkable strategy to deal with this. He always took a large stack of papers and manila folders to meetings. This stack could be as much as a foot tall. When the boss asked him an unexpected question (as happened often to all of us) he would start looking through the stack of paper to assist him with the answer. With any luck, he had the information with him. If he did not, looking through a large stack of paper gave him time to prepare a quality response. In some environments, carrying a tall stack of papers to every meeting might seem unprofessional. In this environment, my coworker had devised a somewhat professional way to stall for time to give him a chance to think.

Likewise, carrying an engineering notebook and writing in it profusely gives an engineer instant respect. For some jobs, acting professionally involves carrying a clipboard or wearing a lab coat. The perception of professionalism comes in many forms. At one company the head of engineering wore his tie in a distinctive fashion. Not too surprisingly, others in the organization mimicked his unusual dress. Those who did not tended to think the mimicry a highly unprofessional display of obsequious behavior. Their opinion did not matter, however. The only opinion that really mattered was that of the boss, and he liked the flattery.

Politics and egos aside, simple office etiquette should be respected. It is surprisingly easy to create a bad perception—or at least to fail to create a good perception—with poor or unconscious behavior. Managing an engineering career is hard enough without seeming unprofessional by ignoring common courtesies. Attention to simple office and social etiquette can be critical. Much of this business etiquette is unspoken. No one tells you; it is not written down anywhere and was not covered in college courses. Worse, different companies have their own cultures and views of proper behavior by personnel of various levels and titles. A short and imperfect list of simple business etiquette might include the following, but your position and corporate culture may emphasize different issues. As always, it is good to listen, observe, and learn:

- Do introductions as appropriate.
- There is usually some sort of executive priority seating at meetings. Be careful to take a seat in a location suited to your level.
- Never disagree with a superior in public.
- Look at and respect executives when they speak.
- Be careful about exposing "dirty laundry" in a public forum.
- Do not attempt to resolve conflicting direction by superiors in a regular meeting. Executive conflicts and confrontations rarely occur in public. Often these are handled in private or by cutting a deal that allows the loser to save face. Most likely you will not be at the meeting where this happens.

Professionalism might be interpreted as behaving like a professional, but most engineers are not professionals in the same sense as doctors, lawyers, and nurses, who have a test that must be taken and passed before they are admitted to the club. While there is a professional engineer test, most electrical and electronic engineers never take it. Perhaps as a result engineers lag behind the other professional societies in pay and recognition. Before someone can write a prescription for a drug, he or she must be a member of the doctor club, but anybody can purchase a compiler, transistor, or integrated circuit. Someone cannot practice law without passing the bar examination, but high school students can be hired by corporations to cheaply hack some code together. One might ask, in the interests of national security, shouldn't access to electronic- and software-development tools require membership in a professional society? Wouldn't that immediately add prestige and force a higher salary for those members of the engineers' club?

Leadership

One absolutely critical factor in continued advancement is being perceived as a leader. Corporate managers are always looking for the rare individuals who take control of an effort and drive it to successful completion. A closer look at these exceptional individuals reveals two types: one pulls the team across the finish line; the other pushes the team. This distinction can be invisible to those not expert in the technology or following the project closely. Pulling the team members to success involves mentoring, parental oversight, and at times personally solving the difficult technical problems. Pushing them, or in some cases beating them, is quite different from the perspective of the engineer. Beaten engineers, for all the reasons already mentioned, ultimately tend to not perform as well as those that are mentored and nurtured. Unfortunately, from a distance, pushers can seem to be the superior leaders because they exhibit the highly prized ability to manage to a schedule. They extract or perhaps demand commitments and then beat the engineers to deliver. Executives in charge get what they want: a visible schedule commitment and visible results. Too often, they also get future customer headaches invisibly accumulated in the mad dash to deliver on schedule.

Managers or leaders to whom getting it right is more important than executing to a schedule are at a tremendous disadvantage in the current world of high-tech development. Their competition for the next promotion will be visible in creating schedules, extracting commitments, and driving the team to meet the commitments. This is visible leadership and is enormously valued by corporations who have time-sensitive marketing campaigns waiting for the released product and new work waiting for engineers to come available. Problems introduced in the mad dash to the finish line will probably not show up until after promotions and bonuses are distributed. Even when issues surface, it will be nearly impossible for those not involved in the end-of-project chaos to understand the true cause.

When problems occur, and they surely will, upwardly mobile engineers respond quickly and forcefully. They commit to delivering a solution on a specific schedule and drive the team to deliver on that commitment. Politically, delivering a commitment to fix an engineering problem is a panacea. However, the reality of engineering can be politically embarrassing. Some things look easy but are later found to have hidden complexities and become a nightmare.

Technically skilled leaders have a continuing tension between demanding that engineers complete a project and in working with them to ensure quality. Quality products provide for the long-term success of the company, but in some companies hitting the schedule provides for the immediate success of the individual in charge.

Grandstanding

Many years ago I was at a bar with some friends when one of them suddenly shouted, "Say anything more bad about this country, and I'll smack you." Now, the ongoing conversation had nothing to do with the country, national politics, or anything else that was relevant to this outburst. Facts were irrelevant when the atmosphere at the bar turned ugly as patrons demanded to know just who was saying bad things about the good old USA. Fortunately, we all made it out of the bar in one piece, but the event made a lasting impression on me. Facts can be irrelevant. Emotions can be manipulated with sound bites, innuendo, and implied accusations.

Engineering sycophants can leverage technical disasters to grab the spotlight and further their cause and political agenda. I have seen people greatly rewarded for stepping in as last-minute saviors even though they had caused the problem months before through either neglect or ineptitude. Unfortunately, corporate engineering rewards high-visibility, last-minute hard work in reacting to problems. Those who anticipate well and create architectures and designs that avoid problems may find they toil in obscurity.

Grandstanding may also occur at the beginning of the project; in this case, it is not the last-minute savior but the interaction between the senior designer and the junior implementer. Some companies have the view that senior people are too valuable to get bogged down in the actual implementation. In these companies, the highly paid person does the architecture and underlying design. The implementation is turned over to more junior employees as the senior engineer moves on to the next project. This is a great theory, but I have repeatedly seen senior engineers design a pile of junk and stick the junior folks with the headache of fixing the mess or wedging patches into an unstable platform. Usually it is not the case that the designers were bad engineers. The problem is that some of this engineering stuff is really complicated and surprises often occur. The surprises must be overcome by the junior folks as the project falls further behind schedule. The other news is that the senior designers escape with their reputations intact because they are no longer associated with the project by the time the problems are discovered.

Stereotypes

People need to have faith in something. This can be especially true of managers in a high-pressure situation. They need to feel as if they can rely on a few critical individuals to take care of problems. In some office environments these go-to individuals become known as the "golden children." The important thing to understand is that gold doesn't tarnish, and neither does the golden child. Once established, these managerial or technical superstars develop a Teflon surface to which stains rarely stick. Much of this is due to simple human nature and the manager's blind and unfailing faith.

A superstar reputation may or may not be well deserved. Indeed, how the reputation is achieved is irrelevant. It is how the reputation is retained that is an interesting and seemingly common aspect of human nature. Unlike actors and actresses who are superstars today and forgotten tomorrow, some managers can have a strong drive to retain the status quo. Busy managers often have little interest in seeking better talent or expanding the skill and versatility of their staffs; they just want the problems to be solved. This is the managerial imperative that promotes pigeonholing engineers as the people who always do certain work.

Tattletales

Every work environment has a mixture of skills and aptitudes. Ideally, the manager assigns work to people capable of doing it in a timely fashion. There are times of resource shortage when the manager knowingly assigns work to an individual that is too hard for or outside his or her expertise. The manager has no other option and simply hopes for the best. There are also times, however, when the manager lacks sufficient knowledge of the subject matter to tell that the assignment is beyond the skill of the assignee. Sometimes the assignee also doesn't know he or she lacks the skill or hides his or her ineptitude and plods along producing nonsense or garbage.

Using your skill and understanding to report in an unsolicited fashion that an individual is not qualified to execute an assignment is rarely well received. There is often an undercurrent of belief by the manager that he or she would know someone is unqualified. This belief can fuel suspicion that you dislike or have something against the individual. Worse, the manager can then be embarrassed if your assertions are later proven correct. You may be trying to help the company or a project, but there is a risk of being too helpful for your own good. You may end up making an enemy of the manager and the unqualified employee.

If supervisors are not knowledgeable enough to see the problem themselves, they won't believe you when you tell them and may ultimately hate you for being right.

Success

Why do some folks advance faster than others? Why do some folks reach higher levels? Are they smarter? Are they better visionaries? Are they better at doublecrossing their competition? While there may be some truth to any of these the most important reason is those who advance are successful at convincing their superiors they should. In contrast, you can impede your advancement by giving your superiors reason to worry about you. No matter how brilliant you are, making those running the company feel uncomfortable most certainly will adversely affect your career.

Anticipate Success

Perhaps the greatest asset engineers can have is the reputation and image of success. Nothing can destroy a promising career as fast as being labeled as negative. Sometimes being a talented engineer can be very bad for career advancement because he or she continually sees reasons that a project might fail or that the expected schedule cannot be met. The politically correct approach seems to be a positive can-do person who doesn't talk about looming problems. The astute engineer then uses his or her skills to overcome project bumps and twists or to get off the project if the problems cannot be avoided or fixed.

Talented engineers can always find areas of technical risk. Unfortunately, many managers and many corporate cultures will not allow engineers to create schedules that anticipate problems and allow time to solve them. At some companies this can be true even in the case of severe problems that are nearly certain to happen. It really doesn't matter if the people in charge don't have the background to understand the problem or simply don't want to hear about it. Such managers and cultures teach engineers to create projects and schedules that anticipate a steady stream of success even if it is unlikely. When the expected problems do materialize, upwardly mobile engineers are presented with the opportunity to become a hero by stepping in to solve the problem with decisive measures and hard work.

Establishing Dominance

Sometimes the most important people demonstrate their position of superiority by arriving last to a meeting and immediately beginning the discussion. This can be very interesting and fun to watch if two people both believe they are the most important. Likewise, the leader may want to be the person who stands at the board and holds the marker. Watching two leaders struggle over one marker can make for an amusing meeting. Engineers can be greatly entertained as the politicians jockey and scheme. Dominant managers may want to hold the meeting in their office at a time convenient for them. Dominance includes demonstrating who is the most important, whose project is more important, whose boss is more important, and any number of ostentatious and trivial interactions. Engineers may avoid career pitfalls by being sensitive to these machinations and thereby minimizing chances of conflict with executives and aspiring executives.

Protecting Turf

If you wrote it, designed it, or invented it, your name should be on it. Some managers try to hide all beneath them. They become the entire corporate face of an organization or project. Managers who behave in this fashion sometimes paint their actions as building a team-first culture. One fairly common instruction from such a manager is, "We don't put individual names on things around here." I learned the hard way that the team leader does not always have the best interest of the team at heart. It is not that these managers steal the work of others; rather, theirs are the only names visible. They refer to their anonymous team, whose specific names degenerate into indistinguishable mush. In this situation executives cannot recognize the achievements of individual members of the team and may even have difficulty naming them.

It seams some humans are simply made from a cloth that readily appropriates the property and ideas of others. The struggle between these and the more honorable folks is primal. Throughout history barbarian marauders haven taken what they wanted from a poorly defended population. Righteousness is no defense, just as it was not for good citizens victimized by thugs over the centuries. Turning the other cheek doesn't work; you just get hit again. We can all wish we lived in a different and fair world, but unfortunately it seems like sometimes we must actively protect our turf from those who would claim it.

Protecting your turf is different from marking it. Marking is what a dog does to a tree or a fire hydrant. An engineer or manager marking his turf is a political gesture that is very different from righteously defending what is yours.

Accidental Success

Over the years I have learned that it is not always correct to attribute success to purposeful intent. There are times when success is a happy accident. Indeed, things are rarely as organized as they seem to be from the outside. The history of successful companies that grew out of hard work and a good idea is much smaller than the number of companies that failed despite expert predictions of success. There seems to be no way for even the most knowledgeable of experts to accurately predict the future success of a business. Some companies are "one-shot-wonders" with their entire history the story of a single successful product. Other companies release a seemingly endless sequence of winners.

A friend of mine was lucky enough to become extremely wealthy as a result of participation in a successful start-up. He shared with me once a remarkable impact this had on his life. The most amazing thing was that people actually listened to him. He was invited to be on corporate boards and to act as an advisor for numerous corporations. True, some of this was driven by the simple desire to associate with and please a wealthy individual, but there was also a sincere desire to learn from this successful man. The joke, my friend said, was on them. There was little he could teach. He had never expected the start-up to be so successful, never expected to become so rich, and had executed no specific plan for any of this.

It seems like human nature to attribute success of a company to the guiding hand of those running it, but human nature may also allow "market forces" to relieve the responsibility of those on watch when a company fails. Indeed, market forces can overtake an unsuspecting company or drive moderately well-run companies to spectacular and unexpected success. One way to look at this is that success is completely random. No one can predict which activities will flourish and which will fail, much less actively drive an effort to success. I find this view distasteful. I prefer to think that success can be purposefully achieved by the hand of a talented individual. My evidence is the occasional resurrection of a mortally wounded company by the injection of new corporate leadership.

If we believe there are people who have the talent to successfully lead a company or project we must then wonder why an individual's success cannot be readily predicted. Real-world experience says that the person selected to replace the leader of a successful company sometimes fails to live up to expectations and that the person selected to manage a project sometimes creates a disaster. There are also times when the engineer chosen to design a subsystem is later found to be unqualified for the job. If there are people who really can do the job, why does eventual success seem so random and unpredictable? One possibility is that the person or group who selected the individual to lead the company, team, project, or design lacked the expertise needed to assess capabilities of the candidates.

In discussions with people I've often heard statements similar to, "We are an important customer so they will put their best person on it." From a more abstract and historical perspective I've heard assertions like, "Sure our society is pretty screwed up right now, but we will solve these problems if it becomes important." Often such statements are followed by a reference to the American on-demand success of World War II, the Manhattan Project, or the Apollo moon landing. The belief is that in times of need the people in charge will select the right person to lead activities and organizations. Poppycock—the implication is that under so-called normal circumstances the people in charge are at best negligent and at worst purposefully allowing inept people to run things. I don't think so. If the people in charge could do better they would. Examples of on-demand success are no more than renditions of historical situations lucky enough to have had the right leaders at the right time. Projects and societies that failed are simply forgotten.

Incumbents

If corporate, national, and world leaders struggle to select qualified individuals, what does that say of your boss? What it says is that he or she is your boss. There is no assurance in this world that your boss will be smarter than you or more qualified than you. He or she is nonetheless your boss and therefore expects you to follow direction. Reality be damned; the boss's title and position state that his or her judgment and negotiating skills are better than yours and implicitly justify a higher salary than yours.

Similarly, junior engineers may be far superior to senior engineers in some relevant technology. It could be the senior people have that position because those running the company trust their judgment. They could also have the position for any number of less favorable reasons. Consider a brief and incomplete list of reasons why people of lesser competency might reach a more senior level:

- They could have been hired into the position with the expectation that they were more expert.
- They could have been an early employee of the company and received the title before there was much competition for the position.
- They were simply in the right place at the right time.
- They may have been promoted out of necessity to the level of a peer with whom they had to work.
- They may have been promoted to an adequate level to be assigned the blame for some problem.
- They could be a relative or a good friend of an important person in the company.

Clearly, there are a large number of reasons that people of less than stellar capabilities might be promoted to a substantial level in the company. There are two things to take away from this discussion. One is that life is not always fair and you might very well be smarter and more qualified than your boss. Two is that he or she is your boss; deal with it and quit whining. Once people reach a particular level they are rarely demoted. To do so would indicate that the previous promotion was a mistake, and few want to admit to a mistake. There is also a fair chance that some amount of incompetence will go unnoticed and nobody knows or cares that you are more qualified than your boss.

Incumbents may also be long-term members of the company and consequently have a very parochial view of corporate cultures. They can be very fixed in their ways of viewing things and doing things and may greatly resist external cross-pollination. Doing things in a different fashion may be far outside their perspective even if the new way is better.

High-level employees may forget the significance of their privileged position within the company. They may be unaware that it takes longer for a low-level person to craft an intelligent-sounding e-mail while their seniors tosses off something with spelling and grammatical errors that barely makes sense. They may also forget that the low-level person has the overhead, delays, and accompanying stress of serving many masters.

Advancement

I used to have a boss who was fond of saying the job is engineering and it is not a popularity contest. Unfortunately, it very much is a popularity contest. It is not popularity with your subordinates or your staff, but with your superiors. Your popularity is paramount when the managers gather and divide up the allocated merit promotion budget. Sometimes following the direction of your supervisor can lead you into conflict with another manager or member of another manager's staff. This can easily result in complaints about your being difficult to work with. It makes no difference that the entire problem was caused at the direction of your boss. You are the person who suffers.

Available money is always finite. When the managers gather during the annual review, decisions must be made on the allocation of funds. Irritating a manager or a member of his or her staff is certain to negatively affect her viewpoint of you. Consensus is a big thing in most organizations, so having an enemy dead set against seeing you advance will prevent a consensus for giving you a promotion or a big raise. Perhaps the most important component to success is avoiding the creation of enemies that oppose your advancement.

Another important factor in advancement is having the right credentials. There are actually two forms of credentials: (1) real credentials, such as meeting the requirement for a college degree; and (2) virtual credentials, where you have to accommodate a requirement existing in the head of the person who must approve the promotion. Lacking real credentials is a disadvantage but is not always fatal. Requirements are corporate policy but can sometimes be waived for favored individuals. Lacking a virtual credential can be more difficult to overcome because the requirement is intangible and can be deeply seated in the psyche of the manager.

The psyche and personality of your immediate manager as well as the general corporate culture can prove a formidable success filter for certain engineering skills and personalities. This reaches to the very soul of the working environment, beyond needing credentials that satisfy a few heartfelt but undocumented requirements. Your managers, and to some extent the corporation as a whole, must determine the type of engineering behavior they want to encourage. Do they want engineers who seek success or those who avoid failure? This is very much a risk–reward proposition. Avoiding failure equates to more mundane and less creative products and designs, but higher risk is inherent in seeking success. Advancement often means staying out of trouble and never overseeing a disaster, and this is the antithesis of taking risk. For this reason, innovative brilliance can actually be a disadvantage since this type of behavior can make risk-adverse corporate leaders uncomfortable.

If you avoid the pitfalls and are lucky enough you may be amazed to find that there can actually be a negative side to getting a promotion. You are now expected to operate at a higher level. What was once exceptional performance can now be viewed as merely satisfactory.

Compensation

Companies often espouse pay for performance; however, there is a fairly recent trend to separate your annual review from your annual raise. You may also hear, "We don't want to make this about money." While supporting pay for performance, it seems as if companies go out of their way to discourage the psychological linking of salary and job satisfaction. This appears to be a real problem for those senior executives who need millions per year to be happy, but for engineers it can be fairly easy, because they tend to enjoy playing with the toys and solving the problems. My experience is that the biggest salary issue for engineers is their desire to be treated fairly. Here salary is used not so much as a way to keep score but as a way to determine who is the better engineer. The difficulty occurs when the engineer's personal view of fairness doesn't match very well with the manager's. Engineers may consider themselves to be smarter and harder working than their fellow engineers who make more money. This leads to discord when the manager disagrees.

A more interesting problem, however, is when the manager agrees. It is very difficult within the corporate structure to correct inequities in engineering pay. I've personally conducted several losing battles to get big raises for talented engineers paid an unfairly low salary. It seems your new salary is always a moderate percentage increase over your old salary regardless of whether you are getting an annual raise or being hired into a new company. Talent seems an unimportant factor in getting ahead quickly. Talent can gradually build a salary at a glacial rate of perhaps 1% or 2% per year faster than the salary increases of mediocre engineers. Salesmanship, charisma, and positioning seem much more important for rapid advancement. As you might expect, this type of thing really doesn't improve the attitude of a talented engineer who wants to be treated fairly.

Written corporate policy at times prohibits raises in excess of a certain percentage. So much importance is placed on your current salary that one might expect some paycheck exaggeration from even the most moral of engineers. Successfully embellishing your salary on even one job change can yield positive results for years to come. The world has established a system where talented, hardworking, and honest engineers can see their earnings fall far behind others who are charismatic, fast talking, and maybe a little shady.

This problem is very hard to solve. Basing pay on an existing number is easy, and anyone with a calculator can do it in 10 seconds. Basing pay on ability requires extended interaction with people who have expertise in the subject matter of interest. These expert evaluators would have to be selected by someone who can identify qualified evaluators. In addition, the evaluators would be risking their reputation every time they endorsed pay for a candidate. Few would want this level of exposure. In the real world, pay for ability and performance is extremely hard to implement. Civilization has gotten to the point where it can severely penalize talented but socially limited folks in all walks of life. Unfortunately, the reverse is also true. Marginally competent but charismatic individuals can rise to very high levels of industry and government.

Success Breeds Success

It can be difficult for engineers to get their first management opportunity. Their superiors must believe they have a chance of success, and management must think they are the right person for the job. Once they have been a manager, however, they can easily be hired into another management position. The next major step is to be the manager of managers. Many organizations will not hire people to manage other managers unless they have already done so. Engineers on the management track habitually have to break through these barriers to remain upwardly mobile in their career. The same is true for senior architects or other key technical specialists. Someone must give junior engineers their first opportunity to lead the development of a subsystem or to set the platform architecture.

There is some interesting psychology at work here. Why not give a talented but unproven person a chance? That would require that decision makers take a risk based on their personal judgment that an inexperienced person can do the job. Hiring or assigning an experienced person exempts one from this responsibility. Doing so merely signifies agreement with someone's previous appraisal that this person has the needed aptitude. The person assigned to the job may not work out, but the decision makers did their due diligence and assigned a veteran who has been there before. Selecting inexperienced people based on someone's personal assessment of their capabilities exposes the decision maker to a great deal of risk.

Risk reduction via the preferential selection of someone who has been there before scales to the very highest levels of industry and beyond. In fact, the higher you go the greater the exposure for picking someone whose experience and qualifications can be challenged. The decision makers have a very clear choice: Avoid failure by picking from a pool of available experienced candidates, or reach for greater success by expanding the selection pool to include talented first-timers. At high levels the exposure is enormous. Avoidance of this risk is perhaps the primary reason for the recycling of mediocre and even bad baseball managers, football coaches, and corporate executives. It is easy to convince yourself that any failings are greatly outweighed by their previous experience.

If you know math, it is easy to tell if people know how to solve a math problem. They might happen to get the wrong answer, but you can look at their work and see that they followed the right steps and had the right idea. If you are a capable engineer, you have a good chance of identifying people who know what they are doing and those who don't. There is generally a well-accepted method of solving a math problem, and there is a correct answer. Engineering is a little more complicated with more variables, but you can look at some designs and know pretty quickly if they have a chance of working or not. Managing projects or people or planning future products is incredibly more complicated. You have to deal with all the complexities of human nature and intrusive and disruptive events of the physical world. Often there is no single right answer. It is a challenge for a gifted and talented person to select a qualified individual, but it becomes literally impossible for someone lacking personal expertise to identify a capable resource to lead a group or task. While selection can be made based on intuition or any number of other imprecise ways, the only mechanism available to minimize the risk of a bad selection is to pick from the pool of people who have already served in the desired capacity. Being CEO of a company, no matter that your tenure resulted in the destruction of the company, qualifies you as a relatively low-risk choice to be CEO of another one.

Once you have served in a certain capacity, you have received a de facto stamp of approval to again serve in that capacity. Aggressive and upwardly mobile people will take advantage of this and will parlay each success into another and another. Success, like no other attribute, breeds more success.

4

Alternate Career Paths

Introduction

Some companies offer dual career paths. This means there is a management path where your career grows by supervising an increasing number of people and a separate path that allows the engineer to remain technical but make more money—up to a limit. Ultimately, the top managers make more money than any engineer, perhaps by as much as two orders of magnitude. The justification for this is that management is a force multiplier. Top-notch individuals leading lesser people can guide them to accomplish far more than the individuals ever could on their own. The theory is that wisdom, judgment, and leadership are the really important issues and are the truly distinguishing characteristics of the most valuable people.

Engineers have the good fortune of a much more flexible career than many other professions. They can stay technical, manage projects or people, or use their engineering background in a broad variety of other ways.

Project Management

After working for a few years as an engineer, some may wish to solve challenges at a higher level. They may wish to see the project in its entirety instead of looking at some lines of code or connecting a few silicon chips. The career path for these folks transitions into program or project management. Different organizations may refer to people who run a project in various ways. I tend to use the terms *project manager* and *program manager* interchangeably, but in some companies the former oversee building a device whereas the latter handle the overall program, which might include marketing, sales, customer service, and the grander aspects of a project that reaches beyond the creation of a physical device. Project and program management can include some or all of the following activities:

- Creating and maintaining project schedules
- Negotiating vendor contracts
- Managing individuals, teams, and vendors
- Gathering and reporting status
- Doing demonstrations and presentations
- Generating, tracking, maintaining, and interpreting project and staff metrics
- Reviewing and prioritizing bugs
- Creating and maintaining project and corporate processes

Project managers need not begin life as an engineer. It is not necessary for them to have a degree in engineering or to have worked as a hands-on engineer. However, it is advantageous for project managers to think like an engineer. An education and background in engineering allows project managers to understand more of what is going on and the consequences of decisions being made. Thinking like an engineer allows project managers to create a logical sequence of steps that yield a completed project. Thinking like an engineer is important if project managers are to detect overlooked tasks and gaps in processes.

First, Do No Harm

Like the medical profession, a good "prime directive" for project managers is to do no harm. Project management exists for the purpose of improving the time to market and quality of the projects. Schedules, statuses, and other project management activities should not be a bureaucracy and an end unto themselves. The gathering of status, maintenance of schedules, and other work should not be so labor intensive that the overhead impedes success. All program management activities need to deal with large enough units of work that the overhead constitutes only a small percentage of the time devoted to that task.

Project managers lacking personal expertise in the project activates may not be able to predict and to avoid problems and may not be able to contribute greatly to the solutions of problems. Though this can be a disadvantage, it is generally not fatal. The important thing is to make sure the project managers do not create more problems.

There are many ways to do harm to a project, and project management run amok can cause damage in more ways than just soaking up engineering hours with status meetings and reports. Excessive management, or micromanagement, sometimes intimidates the developers into undesirable behaviors. Engineers need the freedom to make mistakes, especially when the project involves cutting-edge technology. Reprimanding good engineers who are working hard can lead to excessively conservative designs and strong risk-avoidance behaviors. In extreme cases, engineers may choose to avoid further criticism by waiting for decisions and approvals for every design minutia.

Most importantly, if you keep changing requirements—don't make any excuses—you just don't know what you're doing. Project managers are supposed to be the navigator. Bad things happen when the navigators get lost and keep changing direction.

Failure Is Always an Option

I worked on many projects where the team was told, "Failure is not an option." Unfortunately, some projects proved that false.

One recurrent problem is an entire project based on the unwarranted assumption that every task and activity will be a success. There is no room for error and no backup plan. Always have a backup plan.

Another problem is the inability to confront reality. Why does the schedule keep sliding? The answer, of course, is because you don't have a realistic schedule. It can be that the engineers are incompetent and lazy, but a realistic schedule will take these things into consideration. There can always be unexpected problems, but if the scheduled milestones are missed week after week there is a more fundamental problem. Repeatedly missing milestones is a clear sign that new eyes should be applied to the project. This new person has three tasks: (1) assess the project and determine ways to improve engineering activities so as to stay on schedule; (2) consider the possibility that the schedule is simply not achievable, and deliver a new, realistic schedule; (3) implement (1)'s recommendations. It is critical that the same person who recommends the corrections manages the corrections. I have seen way too many cases where someone recommends impractical ways to hold to the schedule and walks away from the problems to allow the original project manager to continue to take the blame for an impossible situation.

The first-time project manager may be amazed at the number of backseat drivers and advisors who give direction on a project. The higher the project visibility, the more executives will want to help guide the project, be informed of status, and share their know-how and advice. This attention increases exponentially as an important project falls further behind schedule. Unfortunately, too many voices, like too many cooks, can spoil a project.

With support from senior executives, experienced project managers may avert failure by creating a new project that resembles the current project, perhaps with more lofty goals or grander features. Creation of the new project could be justified for any number of reasons such as changing market requirements and the opportunity for additional cost savings. After work begins on the new project the old project is posthumously declared a great success and all are thanked for their contributions.

Elegance, Aesthetics, and Innovation

In most companies, the marketing group, based on its knowledge of customer needs and desires, sets the requirements for new products. Unfortunately, this approach may have the unexpected side effect of missing important features neither the customers nor the marketing folks know lie just over the technical horizon.

Project and corporate cultures can encourage or discourage engineering innovation. Encouraging innovation has the sometimes deserved reputation of generating schedule delays. Unfortunately, discouraging innovation may mean loss of future market share to companies with less risk-averse leadership. It can be difficult to know the ultimate result of a decision to restrain creativity and innovation in the interest of shortening the schedule.

I've experienced several situations where senior folks purposely discourage innovation for fear that such activity will prevent meeting an established schedule. This behavior may even scale to entire corporations. A hallmark of a maturing high-technology company is for the innovative founders to step aside and hand the corporate reins over to "professional management." There is often a simultaneous reprioritization of corporate goals. Professional managers bring a much-needed financial focus on meeting quarterly numbers. Unfortunately, this near-term focus can be at the expense of planning for innovative products 2 and 5 years down the road.

Similar quandaries occur with encouraging elegant and aesthetically pleasing designs, which some believe can be at odds with getting development done quickly. A substantial percentage of managers need to see work activity on a project. They are not comfortable with weeks of design activities that slowly evolve as engineers discuss and argue obscure technical mumbo jumbo. To the uninitiated, planning for elegance is a waste of time. Experienced practitioners, however, have direct observation that elegant designs not only work better but are completed and debugged faster.

Just a Prototype, Just a Demo

It is, unfortunately, a sad fact that many products on the market don't work very well. Even sadder is the fact that some are not really products. They are prototypes or demonstration units not intended to be sold. Some project manager directed the staff to do something quickly and not worry about elegance, error recovery, or supportability. The team rocketed to the finish line, and all were congratulated. Then disaster struck. Customers liked it and bought it. With the marketing folks bringing in wheelbarrows full of money, engineering had no choice but to put some lipstick on the pig and start shipping.

As working engineers know, this happens all the time. Please, Mr. Project Manager, don't do this.

Project Budgets

There are many types of budgets: group budgets, project budgets, capital equipment budgets, overhead, reserves, and many more. The good news is that project managers are not accountants. This narrows the budgetary discussion somewhat but still leaves a number of problems. In large multiyear projects, for example, project managers may have to worry about spending enough money by the end of the fiscal year. There is a very real problem in that spending too little money in one year may result in receiving less money the following year. Many organizations and projects go through an end-of-year buying spree to make sure all the money gets spent. Sophisticated test equipment, powerful computers, conference trips, any and all spare money must be consumed prior to the end of the year. Do people actually do this? Oh yes. Using all the money makes a project manager eligible for a bigger budget next year.

Many corporations have certain times of the year when annual budgets must be prepared and submitted. The deadline can be a little flexible, but at some point project managers simply must turn in their numbers. What if project requirements are not yet known? What if project staffing and needed equipment are not yet known? No matter—the budget must be created and submitted. Project managers can't just make up a number because the folks who process the budgets will reject their input. The right approach is to spend a few hours making several pages of very pretty and persuasive-sounding categories and estimates, most of which are merely speculative. The good news is that the first pass of budget submissions often results in an organization-wide rejection of budgets. Somebody important tells a project manager to cut the $3.7 million budget to $2.1 million. The even better news is that years of experience have taught many corporations the need to go through a mid-year reprojection of the budgets. In a few months a project manager may get a chance to submit a new budget that more accurately reflects reality.

Containment of cost overruns and enforcement of the official budget varies greatly depending on the corporate culture, the visibility of the projects, and the political importance of the project champions. Often more money can be found for important activities, but the offending project manager may be pummeled for the sin of having to ask for more money.

Project Schedules

The first time I was asked to create a schedule I excitedly threw myself into the task. The company was depending on me, and I wasn't going to let it down. I studied some technical issues new to me, took into consideration available resources and their experience, and exerted a great deal of effort arranging and rearranging the order of tasks to ensure they were completed in the most efficient sequence. When I was ready I met with my boss and proudly presented my detailed schedule. In what seemed like only 2 minutes my boss declared that the project was unacceptably long and sent me away to make a shorter schedule. Once more I threw myself into the task. With some tweaking and positive assumptions and with the planned hiring and training of a new employee I was able to reduce the project schedule to only 6 months. This time it didn't even take my boss 2 minutes. He instantly asserted that I hadn't been listening to him when he said he needed a shorter schedule. In frustration I threw my hands in the air and said, "How about 3 months?" He responded, "Good," and the project was begun. No new employee was hired, and about a year and much angst and aggravation later, the project was finished.

Over time this boss became very frustrated about his staff's apparent inability to estimate an accurate schedule. In an effort to understand the root causes (and perhaps somewhat as punishment), he had me document all of my time to a 15-minute accuracy for 2 weeks. At the time I hated this task, but the knowledge gained has served me well ever since. The findings were remarkable. I was a conscientious worker, yet less than half of my time was going into my assigned tasks. Unexpected major consumers of my time included finding out why equipment, components, or software didn't perform as advertised or expected, tracking down lost or "borrowed" test fixtures, responding to spontaneous questions from the marketing group, or handling field support of previously shipped products. It seemed every task had an enormous amount of hidden and generally unsuspected overhead.

Ultimately, the knowledge gained while accurately documenting all my time was most helpful when I ran a small consulting company and had personal control of the estimates used to bid on contracts. After working with my staff for a while we had a pretty good calibration on how long the work would take us and how efficiently we would be able to focus on the actual technical work. We were able to make very accurate estimates. Though nearly every customer complained that the schedules were too long, they loved it when the work was completed on or even before the estimated delivery date. Their joy was reflected in a dramatic percentage of repeat customers.

This success was generally not repeatable while working for other companies. Many companies for which I've worked were more interested in a short schedule than an accurate one. There is a belief, with some truth, that presenting potential customers with a long schedule encourages them to select a competitor who proposes a much shorter schedule. This is a difficult situation because no company can survive if it never wins a contract, yet neither can it survive if it is disastrously late on contracts or creates a bug-infested and poorly performing mess in a desperate attempt to stay on schedule. Schedules can be aggressive, but they must be sane and somewhat feasible or the company should consider walking away from that customer and finding another one.

Even when there is an interest in creating an accurate schedule, the work environment makes replicating my scheduling success difficult as a generic employee. The nature of corporate work generally means working with a variety of new people on every project. This lack of calibration of teammates adds additional unpredictability. Scheduling activities can be much more accurate and predictable if there is familiarity with the team's thinking process and problem-solving abilities.

Another problem is the work itself. My small consulting company was generally hired to do jobs for which we had expert knowledge. We worked on familiar tasks and technologies. It is far easier to predict task duration if you are already skilled in that area. In the general engineering corporate environment, the team must often take on projects that include a great deal of development in unfamiliar territory. Tweaking an old technology is very different from inventing a new one. Unfortunately, it is usually development of the new technology that is high visibility and critical to the success of the company. This creates the problematic scenario of having to invent new science in a fishbowl. This is high stress, high risk, and difficult to predict.

Some things are just very hard to do, and it is impossible to know how hard the next phase is until the current problems are solved. This is known as *peeling the onion*. There are many classic examples of this. In the 1960s, the buzz was that we would have a fusion reactor and therefore access to essentially limitless energy in 10 years. In the 1970s, the same thing was said. At this point, those predictors have learned this is a very hard problem and have mostly given up predicting when it will be solved. Schedule accuracy is sometimes not achievable, especially when there is significant new development or complex technology involved.

Project Status

The goal of a project schedule is to provide an organized framework and sequence of tasks that come together in an optimal fashion. The goal of project status reports is to track progress and to allow for timely intervention to prevent things from going horribly off course. This honorable goal of project status reports may have to be expanded in many directions. The greater the importance of a project and the higher its visibility, the more people feel the need to be informed of its progress. This adds interested observers and other project spectators. I suggest the label *status sinks* for all those interested in the project but who are incapable of meaningfully contributing to the success of the project.

If unchecked, status sinks become *time sinks* and burden the project and project managers by demanding attention to soothe their persistent concerns about bumps and glitches. Their worry burdens the project because they can neither help solve problems nor provide additional resources that can do so. Curiosity is not justification for needing to know the status of a project. The status of a project has no intrinsic value. It is worthless unless something is done with it. Doing something requires making decisions and solving problems. Burdening the team because you are "interested" is inappropriate.

Another status challenge is that some personalities just cannot accept uncertainty. They are genuinely uncomfortable with project managers' saying they don't know when a task will be done or a bug will be fixed. One successful solution I've observed was to give the project status as a weather report—for example, "There is a 60% chance of task 23's being done on Tuesday and an 80% chance of its being done on Friday."

Tricks and Treats

At times, vendors or staff may exaggerate their progress to present a positive image of continued success. Inexperienced project managers may wonder how an individual or company may hope to gain from such exaggeration. Surely they will be found out and will look even worse for having lied. But hope springs eternal. The greatest reason for continuing to present a positive status and to hide delays is the hope that some other group or individual will be forced to step forward and admit a delay first. Another reason is the ongoing hope that some miracle will occur and get their part of the project back on track. Finally, there are congenital liars who just refuse to take responsibility.

For project managers, the issue is not really whether the inaccurate status is the result of ignorance, lying, or somewhere in between. Project managers' job is to ferret out the real situation. Experienced project managers acquire a repertoire of techniques to pierce project obfuscation and to determine real progress. Questions, or lack of them, can often indicate the level of progress of some area of development. When a group responsible for some module is not asking questions, there are two possibilities. One is that everything is obvious and there are no problems and no reason for questions. More likely, however, they have not gotten far enough to actually ask knowledgeable questions. It is not unheard of for longtime program managers to direct errors to be purposefully put into a document or design so that it is evident when someone has started trying to use it.

There are cases where delays and design problems are genuinely unexpected. In one situation with which I am familiar, a vendor was chosen specifically for its years of expertise and success in making a needed subsystem. Several months into the project it became apparent that the vendor would not be able to deliver on its commitment. The obvious question was how could this happen? Hadn't this vendor been doing exactly this kind of work for years? Under direct questioning the vendor admitted that the delays were being caused because it had initiated a major redesign of its product line some time ago. Worse, although the company had a great deal of domain expertise, the people who knew the most were actually in another division and were unavailable. In effect, the company had novices reinventing its corporate technological wheel. What options were available to the project manager? In reality, very few. Months into the project it was too late to look for another vendor.

In this situation, a fundamental flaw in the initial project plan was allowing a vendor to work for months and then be expected to deliver a fully functioning module. Sometimes this is called the *big bang* development approach. The risk with big bang is exactly what was encountered. Either through ignorance or deception the project is surprised with massive delays at the last minute. Far better would be to require finer resolution with multiple milestones where progressively more functional modules are delivered over the course of the intervening months. Even this incurs risk because from time to time individual engineers or companies provide outright junk with no chance of the reputed functionality. Again, the inexperienced project manager may not believe this can happen, because, after all, being caught in this game would destroy the gamester's reputation.

Unfortunately, life is not that simple. Brazen vendors or individual personalities may simply assert that unknowing people just don't know how to use what was given them. Less bold folks may apologize and assert they accidentally sent the wrong module. Sending this "wrong module," however, may buy several more days of needed development. One way to combat this form of subterfuge is to ask that test results of a mutually agreed upon test plan be delivered with the module. Of course, this is additional work, and the vendor may correctly claim that the testing itself is slowing delivery if it is not expressly included in the original development plan.

Here we have a clear choice between the shortest possible development time and a more predictable time. This is a real-world program manager conundrum. The shortest possible development time is almost always the one that is the highest risk. Many times, things can be done to make the development schedule much more predictable, but most such measures seemingly extend the schedule. The executives in charge often decry this apparent extension. Unfortunately, the shortest imaginable schedules usually go awry, and the program manager and developers are blamed for the delays. Heroic extreme measures to recover the delays generally result in further delays, and this vicious and stressful cycle claims yet another project.

Excitement, Responsibility, and Visibility

Project management is where multiple diverse corporate divisions come together and interact. Sometimes the interaction is adversarial and sometimes stressful, but being a project manager is one of the best ways to increase an engineer's visibility and to get perspective on a broad variety of activities within a corporation.

A company I worked for was dealing with some legal issues during the time I was managing the development of a product. In an all-hands meeting the president of the company reported on the progress of the legal activities and said that the hoped-for date of resolving the issues no longer seemed achievable due to the mountain of paperwork that needed to be processed. He guessed at a new completion date but emphasized that there was no guarantee and that it could take much longer. A few seconds later, he presented the ship date for my project and emphasized that we had already missed the original delivery date and that additional delays were unacceptable; we simply must deliver the product on the schedule he'd designated.

After the meeting I approached the president and suggested there seemed to be more than a little irony in this. In the former case, the lawyers and corporate executives have to write, read, and sign a few dozen sheets of paper. In the latter case, the engineers have to figure out why the universe is conspiring against them and invent new science and technology to overcome nature. He was reasonably sympathetic to my point but observed that he had no leverage over the other companies and their lawyers so could really do nothing to speed the process. I observed in turn that I had even less leverage over the universe. Why would he think I could speed the process of overcoming the will of nature? As our conversation continued, we explored the difference in perspective between engineering and business activities in the company. The conversation was far-ranging, nonconfrontational, and nonantagonistic. Unfortunately, we didn't arrive at many answers, but some of the questions were very interesting:

- Why does this seem normal? Why did (apparently) everyone in the company just passively accept that engineers must be held to a schedule while the duration of business activities could not be predicted?

- Was it the case that lawyers and corporate executives were slaving over the documents at 2:00 a.m. while the engineers were trying yet another last-minute bug fix?

- Superficially, the difference was in the acceptability of schedule slips by engineers on technical issues compared with slips by lawyers and executives on business issues. Could the real difference be the prevalent or predominant personalities of the two groups? Could acceptability of the slips be more determined by the assertiveness and dominance of the personality than the issues involved?

- Could it be that more assertive engineers would result in shorter schedules and better quality? If engineers had more of the caustic personalities of lawyers and chief executive officers (CEOs), might they be more resistant to the pressures and directives that caused project chaos and the delivery of poor-quality products?

Not all corporate executives are receptive to the type of discussion described here, so I highly recommend against project managers having such conversations unless they are certain they will be well received by the executive (or they already have another job). Philosophical conversations notwithstanding, the importance of the contribution and the excitement of the job of project management can be among the most exciting. Project managers can be in the center of the storm and can be in the enviable position of having the opportunity to interact with corporate executives and managers from a variety of divisions.

Management

Did you really spend years in college studying engineering to worry about pregnancy benefits and to make sure all employees fill out their time sheets? Did you really invest years in honing your engineering expertise to listen to Tommy complain that Bobby always gets the best assignments? For some people, the answer is yes.

Good technical people do not necessarily make good managers. Likewise, bad technical people do not necessarily make bad managers. Indeed, being a good or bad manager depends greatly on who is doing the assessment and the time span being considered. In the "exact" world of the engineering mind, it may be surprising to learn that there is no universally accepted requirements document by which managers can be evaluated. Indeed, it is not uncommon for the managers' bosses to have an opinion of them that differs greatly from how their staffs feel.

Unlike hands-on engineers who work mostly with inanimate objects, managers bear the burden of having to work with people and make many decisions that have human and emotional consequences. Being a leader of people, however, is the pathway to the highest levels of our society. Military generals, chief executives of corporations, and heads of government all reach the top by being able to marshal their troops, motivate staff, and command the support of their constituents. Indeed, one of the primary points of this book is that most of those at the pinnacle of our society got there by charisma and panache, not knowledge of technology and facts. Engineers interested in career advancement would do well to understand this and to hone their social skills.

Management Training

Companies recognize that most engineers need some amount of training to give them the best chance of success as a manager. I've attended several levels of management training and observed that most of them focus on the mechanics of managing and omit the social aspects. Many management courses implicitly assume that the newly promoted manager understands the need for proper social interaction. The creators of these courses perhaps forget they are dealing with engineers.

Management training courses tend to focus on obligatory administrative activities such as creating objectives and routine supervisory chores such as performance evaluations. Some delve into how to work with various personality types, but I haven't seen any that remind recently promoted engineers that continued promotions likely involve keeping their bosses happy by following their direction. It is just assumed that new managers have the sense to do that. Formal courses excel at communicating the material processes of managing a staff, but the remaining aspects of a leadership position may be left to natural instinct or on-the-job training.

Your technical skill may have been a factor in your being offered a management position. However, nobody can know everything and as you spend more time in management it becomes increasingly difficult to maintain your technical skills. If you continue to be promoted, at some point you will be managing people and projects for which you have no technical expertise. The first and most important point to understand is that you are at risk. Do not fall into the trap of arrogance and insist you would recognize if something was wrong. How could you? If you know little or nothing of the technology involved you are susceptible to misdirection and manipulation by those around you. It's a good bet that any such weakness will be sensed and exploited.

Your Staff

Although all employees represent their company, managers carry a greater share of this responsibility. They are active participants in establishing and maintaining the corporate culture and setting expectations for their staff. The company depends on their management team to hire, train, and get the right employees in the right positions. Assigning the right person to a job is critical because, as they say, you can't get blood from a turnip. Some people are just not very good at certain jobs no matter how hard you push and squeeze and demand.

How do you know if you have the right people for the job? You have to get a calibration on them, and to do so you have to work and interact with them. If you have personal expertise in the subject matter it is pretty easy to know whether staff members know what they are doing. If not, you must use a more indirect approach. Over time, the less capable developers will be constantly surprised when something doesn't work. Everybody makes mistakes, but it is a bad sign if an engineer is repeatedly surprised by problems.

It is the engineer's job to look ahead and think things through. Skilled managers can preemptively address this by asking engineers for a list of things that could go wrong with the project or subsystem. Meaningful engineering tasks always have a large number of things that might go wrong. If engineers can't stream off a significant list of possible problems, they probably don't know what they are doing. However, a manager has little to worry about if engineers walk through a long list of what might go wrong and how their design accommodates each of the problems.

Having the right person for the job today is no guarantee of having the right person for the job tomorrow. The knowledge of a staff ages rapidly. Many technologies understood by engineers today will be obsolete in a few years. Ideal engineering employees know a lot, but they do so because they have a burning desire to learn and understand new things. In many cases, it is far better to have a staff of intellectual athletes than a group of subject-matter experts. Knowledge can become obsolete, but smart and curious lasts the life of the employee. Managing a staff of hardworking, intellectually curious engineers who willingly investigate new technologies is a joy. In such a scenario all that remains is to hire your replacement so you have time to address higher-level tasks and earn your next promotion.

Motivation

A significant part of a manager's job is to create and maintain a positive working environment. Sometimes this includes inspirational and motivational posters, team coffee cups, and T-shirts. Sometimes the same manager blithely ignores or perpetuates indecision and other demotivational behaviors. Indeed, there is nothing quite so demotivating as to work hard all year, be praised for your contributions in your annual review, and then be told, "We can't do much [financially] for you this year."

It seems common sense, but the first step in a manager's successful motivation of employees is to avoid demotivating them. It is also important to understand that employees' perception of motivation and encouragement can very much depend on their individual personalities. Indeed, some personalities feel no loyalty to managers who give them a performance bonus. Many engineers, however, feel some amount of allegiance and responsibility to managers who reward them. To benefit from any resulting loyalty, it should be you and not your boss who hands out bonus checks, stock options, and other rewards. You must be the one your employees look up to, or you will become transparent and insignificant as your staff looks beyond you to those who "really matter."

Marketing, sales, and executive positions tend to structure compensation to motivate specific behaviors and accomplishments. Executives, for example, may be strongly incentivized to drive the price of the corporate stock higher, and salespeople often receive a meaningful portion of their compensation from sales commissions. However, money is seldom used to blatantly motivate engineers. In fact, at several companies I have heard the phrase,

"We don't want to make this about money" in reference to engineering performance and retention.

It seems that some managers and corporate cultures go out of their way to separate the engineering experience from specific monetary incentives. It is quite common to dissuade engineers from focusing on their salaries, instead encouraging them to concentrate on the fun projects and cool equipment. It is relatively common for technical folks to spend their careers with a belief that a visible interest in money is inappropriate. My personal career encountered a paradigm shift while working for a small venture-capital-backed company. It was about a year since the staff last had received raises. Our boss was reporting on a meeting with the board of directors where he told them the staff members were looking forward to their annual salary adjustments. One engineer raised the concern that the board of directors might think we were interested more in money than the work. Our boss explained, "They'd better understand that we are here for money." That brief exchange altered the career of several of the engineers in attendance. Suddenly it became okay for the geeks to want money and their piece of the pie.

For most managers, however, money remains fairly low on the list of ways to motivate a technical person. More popular techniques include playing up the intellectual challenge, the powerful appeal to teamwork (i.e., "Don't let the team down"), and the ever popular approach of beating engineers and demanding they work harder. Saying please and thank you should also not be overlooked.

It can be interesting to observe managers' cycling through various motivational techniques while trying to energize a difficult employee. Different personalities can respond very differently to motivational attempts. Some of the meeker personalities may be highly motivated by threats whereas scheming conspirator-type personalities may care little about letting the team down. On the other hand, aggressive and self-confident people may be highly annoyed and respond badly to demands and threats.

It's not difficult to guess that over time the motivation of choice would be determined by trial and error to be the one that works for most people in the group. This seems to imply that there may be validity to the stereotype of appealing to engineers' sense of teamwork whereas motivating executives requires stock options and multimillion dollar salaries.

Passion

A good manager motivates employees to work hard and to perform well. A great manager ignites passion in the staff. Managers often demand of their staff a commitment in the form of delivering a project or assignment on a particular date. I tend to think a better approach is to demand passion. The real engineering challenge is not to deliver something on a designated date but for the engineers to care about what they are doing and to take personal pride in completion of the assignment. When engineers are "finished" and have delivered on their commitment, questions remain: "Are you proud of this? Is this the best you can do?" Something is wrong if the answer is no.

Encouraging Innovation

Many corporate cultures think they want innovation, but few truly want it. Many managers mouth the words; some mean it; few are ready for it. Wishing for innovation is an excellent example of being careful what you wish for. Innovation involves some risk and a general attitude that can be disturbing to folks who just want their orders followed. Innovative minds can be troublesome to the status quo and established ways of doing things. Before managers encourage innovation they should be certain they are comfortable with the path that follows.

How do you measure and reward innovation, and how do you measure success? How can managers condone innovation—much less encourage it—if it can't be measured? Insecure managers and corporations pose a career hazard for engineers. A single innovative engineer can be attacked and beaten into conformance by the well-meaning but risk-averse corporate management. This can happen over and over, and the company will never know the great things it missed. When the individual engineer is joined by a group of innovative engineers you have a movement. When that movement is supported by the corporation you have a recipe for greatness.

There is nothing unique about corporations. They are made of humans, so corporate lessons can be learned by observation of the greater society. There is a repeated history of advancement and success in nations where citizens are free to challenge, free to protest, and free to solicit ideologies different from those in charge. Likewise, companies that minimize dictating politically correct technology and allow engineers to challenge traditional ways of thinking often see explosive development of products that create and dominate markets.

For a nation or a corporation, the great danger to those in charge is the freedom to propose alternative ideas to a meaningful audience, an audience capable of selecting new leadership. Insecure leadership, or leadership wanting only to have their orders followed, might employ a number of means to suppress this danger to their continued existence. Techniques can range from outright purging of the disruptive influence to appointment of professional fearmongers who shower the working masses with frequent reminders of the grave risks associated with straying from the official course. Those who dare question the course may be called a traitor, may be accused of being unsupportive of the cause, or may even be subject to physical harassment.

A risk-averse company, like a risk-averse society, may institute oppressive agendas that can cripple art, science, medicine, and technology. The innovative mind is the antithesis of rules and regulations. It must be free to think outside the box, to break rules, and to be disrespectful of tradition without looking over its shoulder to see who is watching. Maximum creativity results when the mind is free to roam and to explore without fear of trespassing on

the sacred ground of political correctness and without worrying about second-guessing and meddling by those in charge.

People who do not think in an innovative fashion may trivialize the impact
of random searches, of overseers who review the appropriateness of the
library books you read, and of research findings' being edited to ensure they
suitably represent policy. Those who are comfortable just following orders
may say there is no reason to fear such searches if nothing has been done
wrong. The unimaginative erroneously believe this because they do not
appreciate the true rarity of individuals who ignore possible consequences
and press on with what they believe to be correct.

Most innovators are not so strong. Most suffer genuine damage by simple
exposure to an environment that does not cultivate them. There need not
be specific violations of laws, only exposure to environments that discourage alternative thinking. Furthermore, repressive situations create an atmosphere where accusation by innuendo is tolerated and where you need not
actually do something wrong to get in trouble. You never know when you
may be unexpectedly declared an enemy combatant or a "person of interest"
in a criminal investigation. Your career can be destroyed by mere suggestion
without facing your accusers and without due process of law.

In an oppressive environment—even mildly oppressive—you must be willing to offend those in power and transgress their influence and dominance
to innovate. People willing to do this are rare. The manager truly desirous
of creative innovation must remove obstacles such as even the appearance of
reprisals against those who deviate from the approved and appointed path.
Thomas Jefferson said, "I have sworn upon the altar of God, eternal hostility
against every form of tyranny over the mind of man."

Society as a whole owes a debt of gratitude to those extraordinary folks
to whom freedom and truth are more important than being safe. However,
it seems like far more people have a timid personality: They are uncomfortable without four-way stops and speed bumps and willingly abide by the
fearmonger's dire warnings to stay safely on the approved path. Managers
who desire innovation can support it by emulating Jefferson and removing
obstacles to free thought.

Focus

Unrestrained innovation can doom a company just as easily as no innovation. Excessive rules and regulations should not be replaced with anarchy
and chaos. An engineer simultaneously working on dozens of tasks and
projects is an engineer in chaos. Some managers blithely assign multiple top-
priority tasks to one individual. Should these individuals complain that they
need better direction, the manager often refuses, saying, "You are a senior
person; I expect you to be able to multitask."

There are times when things go wrong, but repeated management indecision and inability to clearly establish priorities are signs of management
ineptitude. It is possible that the ineptitude lies not with the immediate

managers but with their bosses. At some point, the buck can no longer be passed, and some executive must take responsibility and decide what is truly important.

Inability to prioritize and to focus on the development of a small number of products demonstrates a serious lack of self-control on the part of the product planning group. This is the corporate equivalent of an impulsive child with a poor attention span.

I've often heard the excuse that the planners just don't know which product will be successful so several must be pursued. Sorry—it is the job of the product planners to know which will be successful. It is their job to practice due diligence, to look at market trends, and to place bets on a small number of projects most likely to make money for the company. It is unfair and unacceptable to deflect blame onto engineering by betting on dozens of projects in the hope that one might be right. The resulting engineering chaos virtually guarantees failure. In this scenario, however, that failure will be blamed on the engineering group and not the product planners who were unwilling to make a decision.

Knowledgeable managers understand that there is overhead and that productivity is lost every time employees stop or pause one task and begin another. The transition to a new task is called a *context switch* by engineers familiar with real-time operating systems. It is the manager's job to minimize lost efficiency caused by excessive context switching. This means focusing engineers on a small number of tasks and resisting contrary direction from inept or impulsive product planners. Consider these inspirational words from Napoleon Bonaparte:

> Any commander in chief who undertakes to carry out a plan which he considers defective is at fault; he must put forth his reasons, insist on the plan being changed, and finally tender his resignation rather than be the instrument of the army's downfall.

If the engineering manager accepts too many tasks from the product planning group, the engineers will fail, and the engineers will be blamed.

Senior Management

Senior management oversees the formulation and execution of the corporation's goals, objectives, and long-term strategic plans. They decide which opportunities to pursue and which to ignore. Senior management exists to make decisions and to provide direction. In a company of reasonable size, the vice president of engineering is not expected to write code or do timing analysis of hardware. The vice president is there to exercise good judgment, motivate people, and to provide leadership.

Some of the senior management are officers of the company and are authorized to bind the company to contracts. It is important to understand that only officers of the company can legally represent the company. As one of my old bosses used to say, "Your signing a contract means nothing other than a headache for me." Although my signature did not legally bind the company to the contract, later deviation from the intent of the contract would cause a customer relations headache.

More headaches—and corresponding benefits—are part of the job as managers move to higher levels in the company. Obvious changes might be getting a reserved parking spot and the right to approve higher-dollar purchase orders. There are also more subtle changes. One of the most important jobs of senior management is to provide a positive corporate image inside and out. To this end, senior management nearly always supports lower management in public. To do otherwise may create an image of conflict and dissension within the company, and it is not in the best interests of the company to air such dirty laundry. Furthermore, not visibly backing a subordinate opens the possibility that the senior person made an error in hiring or retaining that employee. Senior folks didn't get to their position of power by making mistakes, and they may not be willing to admit one to the public. The desire to hide an error in judgment or misplaced trust may go awry as continued executive support of a bad person creates a worsening corporate image. There may come a time when the offensive person can no longer be retained. Even in these cases the problem is often solved with behind-the-scenes maneuvering and the infamous public statement that the troublemaker "resigned to pursue other opportunities."

Failure to support a lower-level manager who has been loyal can also be construed as a breach of trust by subordinates. Others may be less inclined to remain loyal if their boss develops a reputation of turning against employees during times of stress. Some executives maintain authority by charisma and leadership, but others rely heavily on loyalty and obedience. For the latter, even a hint of betrayal can destroy the executive's ability to command. Continued success of certain executives may require preserving the aura of mutual loyalty at all costs.

Corporate executives deal with problems of a type and severity where exposure of the issues allows criticism no matter the final verdict or the quality of the decision process. In these circumstances facts can be far less relevant than impressions created. This is a world where some engineers might be uncomfortable. It is here that the contributions of charm and social grace can far exceed that of logic and facts.

One thing is clear: The people in charge make a dramatic difference. Be it a sports team or corporation, changes in the people ultimately responsible make a dramatic and at times nearly instantaneous difference. On this issue, there is nowhere to run and nowhere to hide. The people at the top must take the blame for failure and should receive credit for success. It is the people at the top who ultimately determine who gets hired, who gets fired, and who gets promoted. Ultimately, it is the top people who hold all others accountable for their performance—whether things go well or badly, it happens under their watch.

The Stamp of Approval

Engineers able to make the transition from dealing with inanimate objects to working with and directing humans are walking the path that leads to the top of the company. Their first management position gives them a stamp of approval for the next. Their first position of indirect management, as the manager of managers, preapproves them for hiring into even more senior management positions. Their technical expertise provides a background for quality decisions, but it is their social skill that allows them to navigate the political madness that reigns at the top of many corporations.

Technical Consulting

Technical developers are faced with a variety of career options on a regular basis. Commonly available choices include moving into technical management or taking a marketing or sales position (also known as "going over to the dark side"). If the current project is really going badly, other possibilities may include contemplating a variety of forms of suicide, retiring early, or taking a sabbatical. Also popular is the option of changing jobs, and the skills of an engineer can offer ample opportunity to do that.

During times of rapid high-tech growth, job changes may become so prevalent among technical people that even good companies experience substantial staff turnover. There are significant consequences to a high turnover in the engineering staff. These result from the disruption of continuity that occurs when a large number of people leave a project or group. Lessons learned and details of the development process are lost when key members of the staff leave. Good documentation helps but is no substitute for having access to the original developers when there is a question. Likewise, mentoring of the correct development process is disrupted. Newcomers are not indoctrinated in the development methodologies that allowed the company to become successful. During times of economic boom this problem can become quite severe. Continued erosion of the continuity base of some companies could even degrade U.S. national security. In extreme cases, we might anticipate legal action (e.g., laws that discourage frequent changes of jobs) if engineers become too mobile.

Even when the economy is less robust, technical experts are so rare and critical that they often can change jobs at will. Changing jobs may mean doing about the same thing at a new company; the change could also be much more dramatic. One way to significantly change your life is to transition from being a permanent employee to a consultant.

Approaches to Technical Consulting

Once you decide to become a consultant you must then decide your consulting business model. Each model has its own special flavor and lifestyle. Any given lifestyle may be heaven for one engineer and hell for another. This choice should be made carefully and should be consistent with your personality.

One popular consulting approach is going to work for a consulting company as an employee. Working for a consulting company is the closest thing to being a regular permanent employee and is probably the least traumatic way to become a technical consultant. The consulting company finds you work, pays you, and provides benefits. The difference is you may periodically go to work on new projects at a different location. Some find this to be more exciting and more pleasing than working with the same people on the same thing month after month.

At the other end of the spectrum, you could become an independent consultant. As an independent consultant you find your own work and negotiate your own salary. There is no middleman taking a cut of your pay, but there are also no benefits, no paid vacation, and no employment safety net. You live or die on your own technical and marketing skills.

Why Are Consultants Hired?

Technical consultants are hired for two fundamental reasons: (1) because the company simply needs more staff than it currently has; and (2) because the company needs specialized or exotic knowledge. Obviously, being able to provide specialized knowledge to a company is more lucrative than just filling a chair.

There is a noteworthy variant of the specialized knowledge corporate need. Occasionally, management becomes dissatisfied with its technical staff. Development is slow; projects run late and have too many bugs. Managers believe they are providing superior leadership, so the problem must be inferior or lazy engineers. Managers need to hire a top-tier consultant to fix the problem with their staff and to show them how development should be done.

Not too surprisingly, the consultant is usually successful, and this success confirms managers' suspicions that their engineers were not very good. In reality, this is probably not the case. Often the real reason for the consultant's success was his or her focus on a specific task. The consultant was brought into the company with a well-defined task and a well-defined goal. The consultant was generally not distracted because he or she was too expensive to be applied to the more mundane daily tasks. The staff engineers might not be so lucky. They can be redirected on a whim and probably work on several tasks at the same time, many of them poorly defined.

Pros and Cons of Consulting

Why would someone want to become a technical consultant? Common reasons include desiring more money, avoiding a transition into management, or just needing a change. Another possibility is the inability to find a job.

Being a consultant temporarily without assignment is socially more acceptable that simply being unemployed.

Being a consultant has several practical benefits. Some of these are obvious, whereas others are more obscure or even surprising. For example, most people know that the consultant's hourly rate is substantially higher than the equivalent full-time person's. Perhaps a more surprising fact is that some companies are willing to hire self-taught technical folks as consultants, whereas they will only hire college-degreed individuals as permanent employees.

I've worked with several gifted individuals who did not have the opportunity to get a college degree. These people cannot be hired onto the staff of some companies, but they can be hired as highly paid consultants and can be directly involved in contributing to the success of critical projects. Superficially, it doesn't seem very smart of the company that wouldn't hire them to make this arbitrary rule. In a sparse workforce it further reduces the number of available candidates and eliminates some very good people. However, this does make sense when viewed from a slightly higher perspective. Identifying a qualified technical developer is a difficult task for a hiring manager. Some people are very personable and are very good at hiding their lack of knowledge. Brutally speaking, some people are also very good liars. It is hard for a bureaucratic manager to distinguish a slick pretender from someone who possesses genuine talent. Experienced hands-on developers who have moved into management have a little easier time but can still be fooled. Making the arbitrary rule that candidates must have a college degree weeds out people and in theory makes it more likely that the manager will hire a capable individual. There are no guarantees in life, but the theory is that requiring a college degree improves the odds of success.

Advantages and disadvantages of consulting are somewhat open to individual interpretation. Some individuals may consider an activity to be an advantage, whereas others may consider the same thing to be a disadvantage. For example, consultants are often hired to help out during times of extreme stress. They are brought into a firestorm and are expected to hit the ground running. Some folks may like this. Others may consider starting job after job in the middle of a firestorm a significant disadvantage.

It may be better to discuss issues with being a consultant rather than to consider something to be good or bad, advantage or disadvantage. Personal taste and interpretation can be applied to each of the following as needed:

- Staff camaraderie can be different as a consultant. You can be viewed more as an outsider or perhaps as a hired gun. There can also be reduced anticipation of extended employment with the same group, and that may affect the establishment of relationships.

- Consultants can more easily ignore office politics than permanent employees. Many consultants show up, do their work, and go home. Their next raise does not much depend on charming and flattering the boss.

- The company hiring a consultant as staff augmentation generally computes the salary it is willing to pay by simple arithmetic. The consultant presents the company with no overhead for vacation, certain taxes, or health insurance. Determining an exact number can get complicated, but many companies use a simple rule of thumb of about 130% of the equivalent permanent employee.

- However, the consultant offering a specialty expertise is more of a free agent. A presentable consultant with a marketable skill can bypass years of normal incremental raises and corporate hierarchy and get a salary far larger than what would have been possible by remaining as a permanent employee.

- While consultants can bypass normal incremental advancement, a few years as a consultant also interrupts the career process. A hiring manager may be uncertain about where an engineer who tired of being a consultant fits into the standard corporate hierarchy.

- Consultants are occasionally treated as disposable commodities and disrespected. A consultant may get a desk next to a noisy piece of equipment or in a high-traffic area.

- Consultants rarely benefit from bonuses, stock options, and other staff awards and incentives. They are outsiders, not part of the corporate family.

Knowledge Obsolescence

Being a consultant isn't always advantageous, but there is one issue that stands out as a serious disadvantage and needs consideration. Corporate employees are often paid to learn new technology as part of a development effort. Companies are much less willing to pay for the training of a consultant.

Technical expertise has a shelf life, and staying current with technology is a lifelong job for engineers. Consulting engineers may find it difficult to continue selling the same expertise year after year. To remain marketable they may need to learn new technologies on their own time. Learning some technologies can be expensive and require specialized equipment or software. Some consultants avoid this burden by exaggerating their skills in the new area and selling their services to a manager lacking sufficient personal expertise to discern the exaggeration.

Independent Consultants

Independent consultants are a special category that deserves closer attention. They solve technical problems for people but are, of necessity, jacks-of-all-trades. They have limited or no support staff, yet they must market their skills, negotiate the contracts, and do the technical work.

It's a Business

Independent consultants must have saleable expertise and passable marketing skills or they will starve. It also helps to be cautious and to have some business acumen. Consultants can often hear promises of work that never materializes or an insistence on starting the work quickly with the intention of working out contract details later. While engineers may be excited about the work, business executives know that a written purchase order for services is much better than a vague commitment to work things out someday. Finding work is easy. There is plenty of that. The trick is finding work for which you will be paid.

The work of independent consultants starts with marketing their technical skill and then selling it like a salesperson. Only later do the engineers get involved in doing the work. The implication is that the harder the consultants work to market, the harder they work to sell, the more work they have to do, and the more money they make. Working hard as a permanent corporate employee is an investment in the future—a hope that their work will be noticed and that they will be rewarded with a promotion or nice raise. The harder independent consultants work, the more money they make. There is a direct and fairly immediate connection between hard work and more money. For some this becomes an addictive and even destructive behavior.

Headaches Galore

Stress, addictive greed, and the omnipresent specter of starvation are only some of the problems with being an independent consultant. There are numerous legal and contractual issues, and ultimate success requires both technical and social skills.

One of the more insidious challenges independent consultants may face is the difficulty in remaining independent. The legal view of independence can be jeopardized by various situations where a single company provides the majority of their income for an extended period of time. Laws change periodically in this area so they should get an expert current interpretation.

One way to address the issue of lost independence is to work for a company for a legally acceptable number of months, to finish the job, and then to work for another company. Alas, this approach can result in dead time between contracts. If the first company is courteous, it might give the consultant a couple of weeks' notice to start looking for another job. Some companies don't give much notice because they are afraid the consultant will find new work and abandon them while they still have a need.

The customer's fear of abandonment can be generalized into the much bigger problem of required attentiveness and commitment. A company in need of a consultant generally desires the consultant's full attention. It may become very jealous and possessive of the consultant's time. Having multiple simultaneous clients addresses the issue of retaining independence and is a wonderful way for a hardworking consultant to make significantly more money. Unfortunately, servicing multiple clients is in direct conflict with the corporate desire for exclusive attention.

Companies not only want attention; they also want to save money. Many companies resist paying technical consultants by the hour and ask the consultant to work "professional hours"— that is, they ask the consultant to behave like their salaried staff and bill only 40 hours per week while working 50, 60, or 70 hours per week to get the job done. Like most things in life, this is a negotiation, and the consultant and corporation must come to mutual agreement.

Hourly Rate

There has to be a reason to endure all the headaches presented by the independent consultant lifestyle. For many, that reason is a higher salary. This desire for a higher salary should be correctly focused on a higher annual disposable income, not a higher instantaneous hourly rate. Individuals considering a career as an independent consultant should run through the arithmetic to estimate an hourly rate that will give them more money to spend over the course of a year. Factors in this computation include the following:

- The hourly consulting rate is gross pay. No deductions have been taken for federal and state taxes. As a result, consultants must make quarterly estimated tax payments and deal with related accounting issues.

- Consultants get no benefits. They must pay the full cost of their own medical insurance, and every hour of vacation decreases their income. There are no company holidays.

- Consultants must account for overhead support hours. It takes time to create invoices and send them to customers. It takes time to follow up with the customer to make sure they get paid. It takes time to write proposals and to interview with prospective clients. All this decreases the time consultants have available to do profitable work solving problems for paying customers.

- It is likely that consultants will encounter times when they have no billable work. This time could be invested in marketing or perhaps product development, but it generates no immediate income.

Table 4.1 contrasts the source of money to pay for typical work-related expenses and activities. The rows are commonly encountered things that must be paid, and the columns are different ways for which an engineer might work for a company. As an independent consultant, every penny is generated by the billable hours worked. In a corporation, revenue can be generated in a number of ways. Depending on its business, revenue can come from the billable labor of its employees, product sales, profit from investments, sale of assets, government grants, and a bewildering variety of other sources.

TABLE 4.1

Comparative Revenue Sources

Expense	Consultant	Part-Time Employee	Full-Time Employee
Design and Development	Work	Work	Work
Testing	Work	Work	Work
Proposal Preparation	Work	Overhead	Overhead
Accounts Payable and Receivable	Work	Overhead	Overhead
Marketing and Sales	Work	Overhead	Overhead
Business Prospecting	Work	Overhead	Overhead
Business Planning	Work	Overhead	Overhead
Team-Building Activities	Work	Overhead	Overhead
Educational or Industry Courses	Work	Work	Overhead
Medical, Vision, Dental Insurance	Work	Work	Shared
Life and Disability Insurance	Work	Work	Shared
Retirement Plans and Pensions	Work	Work	Shared
Bonuses and Awards	Work	Overhead	Overhead
Social Security	Work	Work	Shared
Medicare	Work	Work	Shared
Federal and State Taxes	Work	Work	Work
Vacation and Holiday Leave	Work	Work	Overhead
Travel Expenses	Work	Overhead	Overhead
Furniture and Office Space	Work	Overhead	Overhead
Property Taxes	Work	Overhead	Overhead
Heat, Lights, Utilities	Work	Overhead	Overhead
Stationery, Pens, Copy Services	Work	Overhead	Overhead
Building Maintenance	Work	Overhead	Overhead
Maid Service	Work	Overhead	Overhead
Telephone Services	Work	Overhead	Overhead
Cell Phones	Work	Overhead	Overhead
Security	Work	Overhead	Overhead

Table 4.1 uses three column notations: Work, Overhead, and Shared:

1. Work: As a consultant, the work generates billable hours. Consultants must furnish their home office, and, depending on their contract, they may even have to pay travel expenses. The design, development, and testing labor of employees of a company result in a product that can be sold. The product can be the labor itself, paper

studies, software, or a physical device. Some of the revenue from product sales goes into a big pool of money called overhead.

2. Overhead: Overhead money pays for many of the corporate expenses not directly related to producing a product. These include buying furniture and keeping the heat and lights on. The only source of money for independent consultants comes from the hours they work. What would have been overhead money comes right out of their paycheck.

3. Shared: These are special expenses or activities whose cost is shared by the employee and the company. For example, the company may subsidize health-care expenses, and it pays a company component of Social Security and Medicare taxes.

The hourly rate of an independent consultant can also be affected by the personality or corporate culture of a customer and even the frailties of human nature. Some customers want the cheapest consultant they can find. To them a $10-per-hour consultant is five times better than a $50-per-hour consultant. It is beyond their perception to consider that the $10-per-hour consultant may take 10 times longer to solve a problem because of either lack of skill or bitterness at having settled for such a low salary. As Red Adair, the great oil-well-fire fighter, said: "If you think it's expensive to hire a professional to do the job, wait until you hire an amateur." There are also customers who want to work only with experts and professionals. For them, the consultant must charge enough to be perceived as a professional. They will not contract with a consultant who charges too low a rate.

Negotiating skill and social skills come into play when presenting the consultant's hourly rate to a customer. Perhaps the hourly rate could be discounted in exchange for a long-term commitment of employment? Perhaps the company could pay some of the money as corporate stock instead of cash? The independent consultant has the ultimate power of deciding to accept a deal or to walk away and look for another client.

Health Care

One big advantage to working for a company is the economy of group health care. The lack of a national health-care system in the United States has the (perhaps) unintentional effect of encouraging many people to work for an established company. This is a double benefit for the existing companies: (1) It provides a large pool of prospective employees, and that tends to hold down salaries; and (2) it tends to help keep the price of their products higher because it reduces the number of people bold enough to leave and start new and competing companies.

There is little doubt that a national health-care system in the United States would create an explosion of entrepreneurial growth as employees who hate their jobs are freed from the restraint of corporate medical plans.

Intellectual Property

Independent consultants are especially well positioned to establish personal ownership of intellectual property. Intellectual property, like copyrights and patents, can create great wealth and long-term income. Many companies put substantial effort into continually growing their portfolio of these assets. Independent consultants can negotiate leniency in the various noncompete and invention agreements that must be signed by permanent employees. This gives consultants the opportunity to develop intellectual property unrelated to their current contract, and really good negotiators may even be able to retain some rights to the work they are being paid to do. Talented consultants can develop a tool or product they can later sell. With some luck, this is how companies are founded and how corporate empires are created.

Starting Your Own Company

Life as an independent consultant is one step away from starting your own company. Potential consulting customers can worry about dealing with a

single individual. They can worry about getting their work done if the consultant gets injured or otherwise distracted from the work. It can be psychologically much more comforting to prospective clients to know they are dealing with a corporation, even when the corporation is predominantly the individual. Independent consultants often take advantage of this psychological benefit and make the leap from being a sole proprietor to starting their own consulting company.

Types of Companies

Companies do many more things than provide technical consultation, but for our purposes there are two basic types of start-up companies: (1) technical services (consulting); and (2) product development. There are also hybrids where a product development company consults to pay the bills until product sales can support the company and where a product company provides consultation on the installation and usage of their products.

Service companies and product companies can have very different approaches to life, and each has advantages and disadvantages. A wise man once told me the problem with a service company is that all the assets go home at night when you turn out the lights. A product company has tangible assets in the intellectual property that goes into making the products and the products themselves. These assets persist beyond the departure of any individual and result in a higher stock valuation if the company should go public.

Service companies sell hours of work. They may bundle the hours into other packages and they may put decorations and trim on the tasks, but ultimately their product is that of billable time. No matter how much the company charges per hour of work, revenue is physically limited to the number of employees multiplied by the number of hours in a day. Product companies have no such limitation. They can generally expand their manufacturing capacity much more quickly and much more economically than the corresponding service companies' can hire billable employees. Product companies' profit is essentially limited by their marketing skill and the desirability of their products.

Product companies can start with an individual or small team and the desire to develop and produce the next great widget. Developing a product takes time, so the company must find a source of revenue to sustain itself until sales reach sufficient profitability. Some divert resources to billable consulting hours; others seek venture capital.

Venture Capital

Venture capital is money provided to a company by investors in exchange for part ownership of the company. Usually the investors want a big part of the company, but the problems run much deeper. Venture capitalists don't give money to just anybody. Many venture-capital-backed companies fail. When that happens, the company that provided the venture capital funds must justify to its investors that it practiced due diligence. It needs to demonstrate that it did everything possible to mitigate the risk. This means that the management staff of the start-up needs an existing track record. The managers of the new company need a suitable stamp of approval before they are deemed worthy to receive the funds. Interestingly, having had an executive position in an established company constitutes most of the stamp. The achievements and quality of the executive's tenure is somewhat less important because charismatic people can usually explain away most disasters that occurred under their watch.

Venture capitalists are to be distinguished from angel investors, who are wealthy people using mostly their own money and therefore are far more likely to make investment decisions based on liking the entrepreneur's thinking process rather than placing heavy emphasis on a prior track record.

One-Man Show

For some venture capitalists, it is the kiss of death if they perceive the leader of a start-up to be a one-man show. He may have achieved spectacular results, but investors can be scared away by the perception that he is not willing to delegate responsibility. When seeking investors, entrepreneurs would do well to explain they are one-man shows out of necessity, not desire. As simple as it sounds, it may not be obvious to investors of great means that entrepreneurs couldn't afford to hire support staff.

Leaders of great technical talent can be so dazzling to the untrained observer that they are at special risk of being perceived by investors as domineering one-man shows. Investors may wonder why it is that leaders have their fingers in all the hardware, software, architecture, and designs. It could be an innocent problem that the little company simply doesn't have the needed resources to attract other top-notch designers. Explaining this to investors lacking personal technical experience can be difficult. In such cases, it may be much more expedient to have a less expert partner interface with the investors while the technical expert stays more behind the scenes, making things work.

Hiding technical expertise from investors behind a management screen is not fair to the expert. In a different world, wizards would be free to bask in the glory of their gift. Unfortunately, most people just don't understand enough about technology to accurately assess someone's level of knowledge or even the criticality of that knowledge to achieving a development goal. More importantly, powerful people of means may not be willing to accept the fact that they lack this needed expertise. It's not fair, but it is the way of the world. Sometimes it is necessary to let the bureaucrats bond with other executives to give technical wizards the opportunity to build new products that may revolutionize the world.

Pizzazz

A friend of mine once interviewed at a little start-up company that had a great location with new furniture and nice offices. He was very impressed and took the job. Later, some criticized the founder of the company for spending too much money on the furniture, artwork, and decorations. The critics said he should have focused more on investing in technology rather than creating an elegant image. Other friends have worked for start-up companies with dingy offices and broken furniture. On occasion, these companies saved money by booking travel arrangements in rundown hotels in questionable neighborhoods.

Furniture, new computers, office space, heat, and lights are expensive, but all of this is insignificant compared with the cost of attempting to create high-tech products with a feeble technical staff. Financial gurus tend to say that it is better to invest money than to save it. Saving money with reused furniture and economical offices may not be the best approach if it creates the impression of amateurism or deters high-quality people from joining the effort. Venture capitalists never want their money wasted, but they understand the need to look like a thriving establishment that is going places.

If you want people to follow you, you have to have some pizzazz; you have to look successful.

Success

What does it mean to launch a successful company? The answer to this question can vary a surprising amount based on the position and perspective of the person asked. The "man on the street" may suggest that a successful company is one that exists for many years and produces a number of products that sell well. A successful company by this definition has turned the corner and has learned what makes a winning product and is frequently able to produce one. These companies often become household names.

The leadership of some companies, however, never quite grasps what it takes to make good products. These people were lucky enough to create a "successful" company by being in the right place at the right time with the right idea but were never able to reproduce that success. They may spend years cranking out market failure after market failure and spending the dwindling revenue of their one success. Eventually, the flagship product becomes obsolete, and they have no choice but to begin laying off people. A company like this becomes a household name only if its collapse is messy and spectacular.

Another, perhaps controversial, view of a successful company is one that makes the founder and his or her friends rich. There are instances where top executives get very rich, yet the company fails to ever produce a product. Even a company such as this can become a household name if the backwash of the corporate collapse destroys a large number of careers and lives.

Indeed, success can mean many things to many people, and engineers would do well to understand the personality and motivation of the corporate founders before joining.

Exit Strategy

Not everyone who starts a company wants to build it into an international powerhouse that lasts for decades or centuries. Some want only to have some fun and make enough money to lead a comfortable life. No matter the goal, entrepreneurial founders should have an exit strategy even as they create the new company. The reason is that the eventual exit strategy can dramatically affect day-to-day decisions.

Actual termination of a company can be surprisingly difficult, encountering delays and complexities in the disposal of assets, the notification of federal and state taxing organizations, and other obligatory activities. However, mechanical issues aside, the real question for the founder is the desired termination method. It is also possible that termination is not the answer. The founder may intend to exit by selling the company to a large corporation or by taking the company public.

Some may find it surprising, but a common motive for a larger company to acquire a smaller one is to obtain the company's technical staff. In effect, the larger company is using acquisition as a personnel recruiting technique. Even if the technical staff is not the primary objective of the acquisition, a quality staff can make the acquisition much more desirable and make it easier for the founder to execute his or her exit strategy.

So Many More

Engineers are smart, versatile, and hardworking. They can be successful at many things, some retaining a connection with their engineering roots and some getting completely out of the field of technology. Clearly, the engineer as a fast food attendant is not especially relevant, but there are a few additional ways to make use of an engineering background.

The Engineer as Applications Engineer

Applications engineers can also be known as field service representatives or technical sales representatives. These are the people who work with the end customers trying to use their company's products. They are on the front line and must confront and overcome all the product warts and flaws exposed by a particular application. This job demands high technical skill but also requires social skills, as applications engineers often interface directly with customers.

The Engineer as Marketer

Engineers with outgoing personalities can be very successful as marketers. They have the technical skills to actually understand what the customer is asking for and to propose a solution that will work. The biggest problem engineers have in marketing is to put their natural pessimistic instincts on hold while they sell the customer with a glorious and rosy picture of success.

The Engineer as Technical Recruiter

An engineering background enables personable and thick-skinned engineers to be fine recruiters. The engineering background allows understanding and communication with both the hiring manager and prospective employees. Thick skin can be necessary since the recruiter can be ignored (or worse) when seeking clients or candidates.

The Engineer as University Professor

Some take this path directly out of college. For others, an experienced engineer returning to an academic environment to pass knowledge on to the next generation seems like the perfect way to wrap up an engineering career. Veteran engineers should consider this option before they retire completely from the workforce.

5

Job Searching and Interviewing

Introduction

Looking for a job is something just about everyone does from time to time. You could be fresh out of school and seeking your first job, or perhaps you've been working for a while and have some good experience. In any case, millions of people are already looking for a job when you join their ranks.

Active and Passive Job Searching

A friend of mine once observed, "The opium drip of my current paycheck continues to sedate me." He hated his job and didn't much care for several of his coworkers, but the continued reception of a paycheck dulled the need to find something new. Each prospective opportunity had to meet minimum criteria before it was given serious consideration. Did the job sound interesting? Was the commute tolerable? Was the salary right? Would it really be any better?

People can passively search for years without finding a new job. Companies can use this inertia to great advantage. Managers giving raises need not please their employees; they need only ensure they don't so irritate them that they transition to actively seeking a job. For some managers, the challenge is doing the absolute minimum their employees will tolerate. They are more concerned about losing head count than about keeping the staff dedicated, motivated, and productive.

Hiring managers, on the other hand, can often sense the difference between those who are passively looking and genuinely interested candidates. It is easy to predict managers' decisions when they are given a choice between eager applicants and those with a casual interest.

When Is It Time to Change Jobs?

Finding a new job goes much faster when you decide it is time to change jobs and subsequently transition to active job searching. Only you know when it time for this transition, but certain patterns can be observed that may influence your decision. For example, if every day you hate getting up and going to work, it might be time to actively look for another job. Other general signs and portents of an impending job change include the following (in no particular order):

- You don't see why anyone would buy the product your team is building. Engineers often have very practical minds and see through the cheerleading and hype to the real product. If you feel strongly that important features are being omitted or if you feel that the product is just plain bad, chances are you are right.

- If every explanation of how the company will make money from a major development effort involves an act of God or the assumed creation and deployment of some new "savior" technology. Fantasizing a future where the infrastructure exists to support profitable operation of your product or service does not constitute a business plan. The plan must be removed from the fantasy domain by enumerating a list of steps with a schedule, costs, and responsible parties that brings the imagined savior technology into existence.

- If purchasers of your company's product or service use it in a fashion that precludes making sufficient profit to sustain the business. This is a problem, but a far worse problem occurs when the company convinces itself that this usage is an uncharacteristic aberration of early adopters. Blindly believing that the problem will go away when larger customer volumes push these atypical users into statistical insignificance can be a recipe for disaster. The disaster occurs when large numbers of customers continue behaving in a fashion that loses money for their company. This is an example of the infamous business plan used by many companies that no longer exist: "We lose a little money on each product, but we'll make it up in volume."

- The company has changed its logo multiple times in the last few years, which hints that the executives may not be focused on the really important issues.
- Your boss asks you to hang on for another few months without pay.
- People you hate or who hate you get promoted to executive roles in the company.
- You are not learning anything or having fun.
- The people in charge keep hiring people even more stupid than they are.
- Your company has been shrinking in market share and number of employees for several years.
- Talented engineers are repeatedly stymied by less creative and capable individuals—individuals who seem to fear innovation and go out of their way to block it.
- A large number of managers and corporate leaders are more interested in not failing than in leading and succeeding.
- Your manager thinks the only way for you to prove your worth and commitment to the company is by working long hours. Creating a culture of working hard for the sake of the appearance of working hard is counterproductive. Good engineering is about efficiently solving problems and building tools to enable greater productivity. Good engineering is more about working smart. When the culture overemphasizes the image of hard work, the staff responds by working on problems instead of solving problems. Too many years of this will doom an organization due to low productivity compared with the competition.
- The company has decided it needs to do more with less.
- You received an exciting offer for a new job.
- You can retire.

There may also be warning signs associated with your immediate supervisor:

- Your boss believes that managing amounts to gathering status and propagating it up the management chain.

- A corollary to the previous point is that as a project falls further behind schedule the manager has status meetings more often.

- Your manager believes that the secret to successful management is to write down all the commitments and milestones. Writing things down is a good idea. This becomes a problem when the list only provides a written record of broken commitments and missed milestones. The real secret is in actually doing things and doing them well. The benefits of a written list are quickly lost if the labor associated with maintaining this list detracts from doing real work.

- The manager rarely suggests a solution to a problem that actually works.

- The manager pushes people to solve problems beyond their capability. This becomes dramatically worse if the manager cannot tell when a task is beyond the employee's ability to solve.

Layoffs

One excellent reason to look for a new job is when you've been released from your old job. This happens quite often in the modern era of outsourced labor, downsizing, rightsizing, and annual CEO salaries approaching a billion dollars.

There are a number of reasons that you might be laid off. Development of a core technology may fail, product sales may fall short of expectations, or your job may have been sent to a foreign country where labor costs a small fraction of your pay.

Outsourcing

The last few years have seen a marked increase in layoffs caused by outsourcing of jobs. Outsourcing is the process of sending jobs to a cheaper labor environment. It is especially onerous because it not only inconveniences those who have lost their jobs but also does long-term damage to the underlying U.S. economic framework. The first companies to outsource work make out like bandits, and to some extent they are. These infrastructure parasites reap huge rewards due to the benefit of their lower cost labor, but they do this by raiding the national economic infrastructure. Competition forces first hundreds, then thousands, of corporations to outsource an ever increasing amount of work. The working masses are adversely impacted when thousands of corporations remove jobs from the national environment. At some point, too few people retain jobs to buy the products now being made so cheaply. Outsourcing is no more than a ponzi scheme that, when carried to its logical conclusion, destroys the foundation of the U.S. economic system.

In World War II the president of the United States went to the major automobile manufacturers and asked them to start making ships, planes, and tanks. They made a lot of them, and that was a major factor in the Americans' winning the war. After decades of certain tax structures, wages, liability issues, and other financial demands and constraints, the president can no longer do that. Well, to be more accurate, the president can ask, but the automobile manufacturers will have to get in touch with their foreign subsidiaries and partners to see if they can change production as requested.

By far the biggest problem with outsourcing is that it often leads to the American consumer's subsidizing technological and manufacturing research and development that will enable foreign countries to become formidable competitors. One could predict that it is only a matter of time until the continued outflow of skills, technology, and jobs endangers national security.

Severance Packages and Retention Bonuses

Regardless of whether you were laid off because of outsourcing or some other factor, the layoff can sometimes be a gray cloud with a silver lining. Reputable companies often ease the pain of being dismissed by providing the employee with a severance package. The change in control caused by a corporate acquisition can yield an especially generous severance package. These packages sometimes vest employees in their 401Ks, company stock, retirement plans, and other assets. Sometimes they even pay health insurance for a few months.

A corporate acquisition brings many good things, but it can also be disruptive. It is not unusual for numerous high-level personnel to be replaced during a takeover. These new folks may not know much about the business; however, they are known to be loyal to the new owners, who can be confident that their interests are being well represented in every decision. Unfortunately, the existing staff can become agitated and restless in the time it takes the newly installed leadership to acclimate to their latest responsibilities.

Layoffs, new leadership, and related uncertainty can be very stressful to the staff. While the company may want to shed some people, it must be careful not to cause a stampede. Massive layoffs can lead to bad morale and widespread concern about job security. The company is doomed if too many important people run for the door. The traditional way to handle this is to establish retention bonuses for critical employees. Staff engineers are generally considered replaceable commodities and don't often participate in retention bonus programs.

What Kind of Job Do You Want?

The first step in searching for a new job is to understand that you are really looking for a better one, not just a different one. The question then becomes what constitutes a better job? More money? Greater responsibility? Less responsibility? Perhaps a shorter commute? Determining what makes a better job involves two components: (1) identifying what needs to be removed from the workplace environment to improve your job satisfaction; and (2) determining what is missing from your current position. Once you make these determinations and know what you want, finding it becomes much less challenging.

It is interesting that one of the easier things to do for an engineer changing jobs is to make more money. Some managers and corporate cultures pay employees just enough to keep them from leaving but willingly pay more to attract new and unproven talent. This psychology creates a market where engineers can raise their salary with a few job changes. There are negatives to this tactic:

- One must be careful to avoid being labeled a job hopper. Being so labeled can adversely affect future prospects.
- Frequently changing jobs may prevent vesting in pension plans, stock options, and other lucrative benefits.
- One may accidentally jump from the frying pan into the fire. If there are some things you really hate about your current job, you owe it to yourself to make sure they are absent from the new one.

Temporary, Permanent, or Part Time

Being a part-time employee can be ideal for an engineer raising a family or with other special demands. Unfortunately, the majority of engineering companies need individuals to solve problems on a predictable schedule where the amount of work required may be poorly understood. This means engineers must adapt to an unexpectedly hard problem by working long hours—which are not very compatible with being a part-time employee.

Temporary engineering positions exist but are more interesting to summer interns or consultants than to someone seeking to establish a career as an engineer. Some companies may want to hire a new employee on a temporary-to-permanent status; that is, the employees join the company with no mutual guarantee of commitment. This is something of a test run to see if both parties like each other. This approach doesn't make much sense to me for several reasons:

- As the hiring manager, I would be looking for people who liked what they saw and heard at the job interview and want to commit to my company. I would be less interested in candidates who felt they needed the additional exposure of a temporary position until they could make a decision.
- As the prospective employee, the same is true. I would be more interested in a company that believed I was the right fit for their future plans and committed to hiring me.
- Occasionally, a company tries to hire a consultant as a temporary-to-permanent employee. This doesn't seem to make much sense if the

consultant wants to remain a consultant and makes no sense at all (to the consultant) if this is the company's attempt to hire him or her at a discounted hourly rate.

- Finally, the modern reality is that little mutual commitment exists even for employees hired with the intention of making them a permanent addition to the team. I've seen cases where the employee never showed up for work or quit after just a few days. I've also seen cases where the employer released a perfectly fine new addition after only a few months because the corporate plans changed.

Joining a Start-Up

A start-up is a small, privately held company unable to support itself with sales but with a plan to become very successful. In such an environment, the company executes a purposeful strategy of deficit spending that requires an influx of money from investors. This situation can persist for several years until the company can support itself or the investors get tired of providing money. This presents a high-risk environment with the possibility of high return for the right team.

There are negatives to a company purposefully overspending its revenue. Employees on corporate travel may be expected to sleep in fleabag hotels and make do with damaged office furniture. The company may have few if any development processes, and long hours of work may be the norm. The good news is that a start-up company may offer incentives to its employees. Often this incentive takes the form of the right to buy stock in the company at a bargain price. The "bargain price" is computed working with the investors and corporate financial experts. The idea is that someday the company will be a successful and the employees can sell their stock for an enormous profit.

The future is uncertain when trying to predict the value of the stock. Nobody knows for sure how successful the company will be, so nobody really knows how much the stock will be worth. Crystal balls are consulted, polished, then consulted again. Ultimately, consensus builds to a value that employees and investors like. Reputable start-ups explain to prospective employees that the anticipated stock value is only an estimate and cannot be guaranteed. They may also be willing to disclose the approximate number of shares allocated to certain employees or positions.

The hard work and high-risk lifestyle of a start-up is not for everyone. Worse, most start-ups fail, so diligent consideration of the business plan and its likelihood of success are needed to avoid investing time in a doomed enterprise. Fortunately, many reputable companies are willing to share their plan for success prior to hiring. Unwarranted exuberance is the biggest problem. Beware of companies that create a business plan where a miraculous technical advancement makes the money-losing venture profitable. Plausible business plans lay down a funded and scheduled strategy for all needed technology. To be fair, miraculous savior technologies are fantasized in companies of all sizes. The difference is that a large company can overcome the miracle's failure to materialize by subsidizing the effort with profits from other ventures.

Joining a Company in Transition

Companies in transition are those experiencing growing pains as they evolve from little start-ups to successful medium-sized corporations. In many ways these can be the most difficult yet most emotionally rewarding companies for which to work. Emotional rewards are earned by overcoming the many obstacles, getting good processes in place, and sometimes just surviving another day.

Chaos may dominate the landscape on a regular basis as the growing company tries to stretch its wings and fly. The company may have little restraint and may chase every opportunity, heavily stressing the engineering group.

Sometimes the original founders are still with the company and are struggling to adapt to the growth they have overseen. The founders may have personally experienced success using technology and techniques no longer relevant to the larger projects the company now undertakes. Alternatively, the founders may overcompensate and force crippling process and documentation into every project. Success might take years and be quite painful, but at the end of the journey engineers may find a nice position in a publicly traded successful company.

Joining a Mature Company

A mature company has it all. It has enough money to pay good salaries with nice bonuses, and it can afford premium travel accommodations and comfortable offices. It likely has reasonable processes in place and knows where it is going in the marketplace. It can offer a great deal of security, but you probably won't get rich.

Marketing Yourself

I'm aware of a case of relentless self-promotion by an engineer. He regularly placed notices of "brown-bag" lunch training sessions around the office building, in elevators, and on bulletin boards. He would put together short sessions on a variety of topics and present them. It wasn't long before he'd successfully created the impression that he was an expert in technical, managerial, and operational aspects of the company. Soon he was viewed as a corporate treasure and rapidly advanced. As of this writing he is a senior

executive of a major corporation. His aggression, political savvy, and lack of modesty served him well.

Becoming Well Known

Opportunity for advancement increases with every additional corporation and hiring manager who knows you or knows of you. Reserved or shy engineers can take heart that brazen self-promotion is not required. You can become well known with such mundane means as conference presentations, magazine articles, or a few patents. Moreover, your resume can spread your "fame" far beyond the hiring managers who attended your conference presentation or read your magazine article.

Several forms of self-promotion may actually pay engineers a token sum for their efforts. Patents are a little different in that applying for a patent costs a finite amount of money. Employees of a small company may therefore be at some disadvantage. Discretionary spending for patents may not be a priority for a small company. When a smaller company does spend money on a patent it is usually based on an idea of one of the founders or executives and is viewed as critical to the company's future. Larger companies have more money to spend and routinely pursue patents in more areas.

A few words of caution about self-promotion: Most companies don't like their staff engineers to be well known. They are well aware that highly visible engineers cost more to retain. Some companies allow only their senior executives to be particularly visible.

Networking

If you are hardworking and competent, the best way to find a new job is through personal references. Hiring managers are much more likely to hire someone who is brought to their attention by a mutual friend. The value of networking is well established. It pays to stay in touch with former coworkers and business contacts.

Working with a Recruiter

Some of the best jobs are not known to the public. Getting your resume into the hands of a few recruiters can vastly improve your chances of being considered for one of these elite jobs. Furthermore, good recruiters are likely to know far more hiring managers than the typical engineer and can be far more adept at getting your resume in front of the right people.

Extremely versatile people may run into conflict with some recruiters. A friend of mine is an experienced engineer capable of both hands-on hardware and software engineering. He has also worked for several years as a manager, he ran his own company for a while, and even served as a system architect for major corporations. Over the years, multiple recruiters have lectured him to decide what he really wants to do and pick a specialty. These recruiters appeared incapable of conceiving of an individual who could do all these things well and had no idea whatsoever how to sell such a person to a hiring manager. My friend found out that these recruiters viewed his talent and versatility as a burden rather than seeing it as a significant real-world advantage.

Perhaps some of the perception of being a burden was merely the recruiters' echoing the concerns of hiring managers who wanted only to hire an engineer and not a renaissance man. Many hiring managers are looking for a person with a number of years of experience with a specific expertise and aren't interested in trying to understand how a broader background might be helpful.

Applying for a Job

If you are serious about getting a new job you must continue to apply until you have a written offer. Never stop after a promising interview. Don't stop even if you are promised a job verbally. I can't count the number of friends who have been promised jobs only to find out funding failed to come through or someone failed to approve the hiring.

Advertised Jobs

Sometimes job openings are not advertised, and candidates are interviewed secretly behind the scenes. This can be for any number of reasons, including because the company wishes to replace someone and wants to find his or her replacement before the incumbent is released. Most of the time, however, jobs are publicly advertised. Interested parties must look for these advertisements in the right place and submit an appropriate application.

Looking in the right place can include automated electronic searches; however, these searches are not perfect, and you should not rely exclusively on them. My resume is very clearly associated with the engineering profession, yet one of the major online employment sites matched me with a job as a shoe sales associate in a department store.

An advertised position might say to submit your resume to "Sarah" or "Jorge" or some other specific name. Often this person doesn't exist. The name simply serves as a code word to tell the company how you learned of the job.

Some care may be indicated when blindly applying for an advertised job. You may actually be applying to your current company, perhaps even for your current position. It is also possible the human resources department responsible for processing your application may not forward your resume to the hiring manager because the reader didn't see certain expected key phrases or was too busy to process it.

Resume Shopping

Not all advertised jobs are for real positions. Some advertisements are intended only to collect resumes that will be used to bid on a contract. It is even possible that numerous companies advertise for resumes in pursuit of the same contract. Occasionally, the identical resume is bid by multiple companies. Unfortunately, the owner of the resume is not guaranteed to be hired even when his or her resume helped win the contract for the company. Unless the resource was listed as a key implementer the company can substitute another, perhaps cheaper and less qualified resource.

The Interview

Luck has smiled on you, and you have been asked to come to the company for an onsite interview. Experience says that many companies don't do a very good job of executing the actual interview process. You can help them be a little more organized, and in the process help yourself, by asking a few questions:

- Ask for an interview itinerary or agenda. It should have the names and titles of the interviewers. In some companies the title may not convey the level or actual importance of the individual, and it is reasonable to ask for clarification. This information will tell you a great deal about the company and whether it correctly understood the position you were seeking. For example, if you expected to interview for the position of director of engineering, it is a really bad sign if the people interviewing you have the title of engineer or even senior engineer. All positions should be interviewed by peers and superiors because lower-level people may be coveting the same position as you. I've seen many companies that operate with the presumption that interviewers are altruistic and will recommend a formidable and talented future competitor for advancement. These companies are naive and wrong. It is a rare person who endangers his or her advancement to do what is right for the company.

- Ask for some friendly advice on what kind of clothes would make the best impression on those interviewing you. Sometimes a suit is not the right answer for a technical interview.

- Ask how reimbursement is handled for incidental expenses. I can't speak for other occupations, but an engineer never pays to go to a job interview. A company serious about hiring you will pay all reasonable expenses associated with your interview.

- Ask for a couple of names and phone numbers so you have someone to contact in case something goes horribly wrong on interview day.

Proficiency Tests

Some companies feel strongly about proficiency tests and others don't. Requiring a candidate to pass a technical test can make sense for a number of reasons. It could be that the company has so many applicants that it needs some way to prefilter the candidates. True, a few good people may be shed along the way, but the company really doesn't care because it has so many applicants.

Proficiency tests can come in many flavors: a written exam, "thought" questions having no set answer, or an interactive session at a whiteboard with an interviewer.

Other Tests

I went to a job interview once where the human resources person gave me a word association test. I didn't especially see the relevance of this, so I slowly stepped through song lyrics as the interviewer presented words. Whatever lyric word was next I provided as the answer. It was a lot of fun, and the interviewer looked very confused to the point of distress. After several bizarre responses, I told him what I was doing and started laughing. I went on to explain that I thought his word association test was silly and a waste of time for both of us. Of course I didn't get the job, but I also didn't want to work for a company that did things like that.

No doubt the company had good intentions in doing psychological screening of its applicants. On the other hand, half the talented technical people I know are more than a little nuts. It's really hard—perhaps impossible—to get great creativity without some unusual psychological components. Those who design or approve these psychological tests want to pound the employees into a preconceived mold. They know what they want, but what they want may not be best for the company in the long run. In fact, it is nearly certain that a staff consisting entirely of milquetoast team players who follow the rules will eventually doom the company. Sometimes it takes a few activists to stir the pot and expose and overcome serious problems.

Rudeness

I've heard of several instances where the candidate was made to wait a long time—in some cases more than an hour. This is either a sign of severe disorganization or a dominance display. Either way, you may want to rethink taking a job with a company that behaves in this fashion.

I'm also aware of a case where a senior individual endured a rather brutal 2-day interview process at a major corporation. After speaking with numerous people he was escorted to the lobby to await a meeting with the division vice president. This was the final scheduled interview. About 15 minutes after the planned meeting time he walked up to the receptionist and asked when he might expect the vice president. The receptionist called a few people and then asked him to have a seat. A few minutes later a rather frightened

secretary came and told him he needed to return to the human resources building. She nervously declined to answer any questions. Back at human resources he found that the interview was over and that he was not chosen for the job.

The candidate was most unhappy with his treatment and wrote a letter of complaint after returning home. He was disappointed about not being chosen for the job, but not angry. His irritation centered on the rudeness the vice president had displayed. The candidate had traveled across the country for the interview and the vice president hadn't even taken the time to thank him for coming and to personally deliver the news that he was not selected. Instead, he was left sitting in the lobby and could have been there for hours had he not approached the receptionist. A footnote to this story is that the company instituted interview training for its employees as a result of the candidate's letter of complaint. Amazingly, this training was conducted by the division vice president—the very person who had caused the problem.

Fear of Talent

Not every manager wants to hire superstars—even if the superstar is economical. Sometimes this comes from a genuine concern about keeping the individual challenged and interested in the work. Unfortunately, there are also times when the concern is about bringing into the company someone good enough to threaten the security of those doing the interviewing.

Seasoning

Discrimination based on age is, of course, illegal. It exists nonetheless. Gray-haired individuals may be routinely asked how long they plan to continue working. On the other hand, discrimination against younger folks also exists. Employers may look specifically for individuals who have some "seasoning" or have "been around the block." These employers are specifically looking for experienced individuals who have learned from their mistakes and can mentor others to avoid those mistakes.

Typical Interview Questions

Job interviewing can be very stressful for some people. This stress can be reduced by early preparation for the interview, anticipating likely scenarios, and having ready answers to common questions. Numerous books exist that can tell you how to sit and whether to lean forward if the interviewer leans forward. You would also do well to have the mental discipline to respond well to unexpected questions. Finally, remember that some interviews are made purposely stressful to determine how you react in difficult situations. As the stress level increases some people talk more and faster. It can be important to have the skill to know when to stop talking. Silence is not necessarily a bad thing.

Why Do You Want to Leave Your Current Job?

Be professional. Take the high road. Good reasons to leave a job are to advance and to seek new challenges. Bad reasons include a pending sexual harassment lawsuit, that your current boss is a jerk, or that the company's management consists of clueless buffoons.

Why Do You Want to Work Here?

Ideally, you have done your homework and actually know what the company does and what it wants to do in the future. Here is your chance to explain to the interviewer not only how your skill and talent can help the company today but also how you can be instrumental in getting the company to its future goals.

What Is Your Greatest Strength?

This is a version of the question, "Why should we hire you?" This is a great opportunity to tell the interviewer about your most significant achievements. Many interview candidates ask for the chance to do something. You will distinguish yourself from the crowd by pointing out what you have already done and how you will use that experience in the new job.

What Is Your Greatest Weakness?

Do not be surprised by this question. Many, many interviewers ask it, so have a good answer ready. The tone of your answer can somewhat depend on your age. If you are young in your career, you can say you are still learning your strengths and weaknesses. If you are older you can pick something and explain how you worked hard and corrected that weakness. The most difficult situation is for those of you in mid-career. Here it may be best to pick one of your many weaknesses and explain the steps you are taking to overcome it.

Tell Me about Yourself

The interviewer asks this question to get to know you and determine if you will fit into the prevailing corporate culture. As you don't work there, you most likely don't know much about this culture. Therefore, you can't make up an answer the interviewer will like. This is a good thing because there is more to interviewing than getting a job. You really want a job you enjoy, and this means you fit in with the current employees and culture. So—tell the interviewer about yourself. Tell the truth. If you won't fit in, it is better to learn that now rather than after you accept the job.

Other variants of this question include, "What do you do for fun?" and "Why did you become an engineer?"

Tell Me about a Difficult Problem You Solved

When I ask this question I want to know about a technical problem or implementation bug the interviewee tracked down and fixed. I want to see his or her thinking process. Others may want a description of your resolution of interpersonal problems. Have some good answers to this type of question; this is a chance to separate yourself from the competition.

What Are Your Salary Expectations?

For most people this is a fairly easy question. You expect to get a modest increase over your current salary. For those of you who believe you are a skilled negotiator, here is your chance to shine. Most experts say to avoid salary specifics prior to the actual offer, but this seems a little silly. The interview is just a waste of everyone's time if you expect twice the salary the company is willing to pay. The most important thing is to know how much money it would take for you to happily accept the position. If you know that number, what is the harm in offering it?

Negotiating the Offer

The secret to negotiating is to be willing to risk losing. This is true of all negotiations whether you are buying a car or a house or are changing jobs. As soon as you decide you absolutely must have a particular car or house, you are at a disadvantage in negotiations. The same is true with employment. If you have no job and really need one you can't be too picky or push the negotiations too far. As the experts say, the best time to look for another job is when you already have one. The experts also tell you to get the offer in writing before you make a final decision or quit your current position.

Know What You Want

When you arrive at the interview have an idea of what it will take for you to accept the job. Make sure you know what it will take for you to accept the job when you leave the interview. How much did you like the work? How much did you like your boss? Does it look like there is a good chance for advancement? Is the commute acceptable? Knowing what you want is critical but can be very hard to determine. Talk with friends, your spouse, or yourself, but figure it out. If the company's first offer is acceptable, jump on it.

Avoid the Low Bidder

Ideally, you are selling your talent to a potential employer. Your approach should be to explain why your skills are advantageous to the company and will help them for many years to come. Some employers, however, are looking only for the cheapest labor they can find. For a number of reasons you should avoid such companies. In general, you benefit greatly by presenting yourself as desirable because of talent and unique skills, not because you are cheap.

The Performance Matrix

When you begin work at a company you do so with a title and level likely to establish your peer group for many years to come. Companies more interested in talent than economy may nevertheless be willing to pay a little more to add new blood to their staff. This gives you a nice starting salary but may have the unfortunate long-term consequence of smaller annual raises until your salary falls back closer to the median for the peer group.

Many companies use a performance matrix to determine their employees' annual raises. This matrix consists of two components: engineers' performance and the amount engineers' salaries deviate from the median. This matrix approach allows engineers receiving a meager salary to get a big raise even if their performance is only mediocre. Similarly, engineers who perform very well would get a bigger raise unless they are already being well compensated.

This approach has the arithmetic result of pushing most employees toward the median salary. The exception occurs when a high-performance member of the peer group gets promoted into the next group. A newly hired employee, however, may not be eligible for advancement the first year or two since corporate rules often limit the frequency of promotions, and being hired usually counts as a promotion.

Your Level

People like to get regular promotions. Instead of saying no to employees who ask for one, it could improve morale by just giving them a small promotion. This means the corporation must have a multitude of levels within which employees can be promoted. Employees hired as a senior engineer may be surprised to find themselves eight or more levels below the president of a large company.

Titles can be misleading. A senior manager sounds important, but the corporate hierarchy may include associate directors, directors, managing directors, junior vice presidents, and so on. Before accepting an offer, it may be beneficial to ask for a description of the corporate structure. A reputable company will provide this.

The Fine Print

The engineer's total compensation may include several components:

- The company may offer a job candidate a hiring bonus. This bonus is often paid the first payday but must usually be repaid if the employee leaves the company within the first 12 months.

- A regional cost-of-living adjustment if the candidate must move to an area with a more expensive cost of living.

- Relocation expenses if the candidate must move a substantial distance. Covered expenses vary from company to company but can include payment for a moving company to pack up and move the household, purchase of the house, transportation of family automobiles, temporary use of a rental car, a month or more of housing expenses (the duration may be negotiable), and payment for incidental expenses.

- Stock options or warrants may be part of a comprehensive hiring package. This offer often comes with a vesting period, meaning you lose some or all of your rights if you leave the company before a predetermined time. The actual vesting can operate in many ways to accommodate different corporate goals and taxation issues.

- An annual performance bonus might be discussed but is usually not guaranteed. Determination of the actual bonus can also be quite complicated and might include personal performance, performance of your corporate division, and the profitability of the company as a whole. There can also be a waiting period before you are eligible for this bonus. New hires may not be eligible their first year of employment.

- Matching funds for a 401(k) retirement plan. The matching percentage has been declining for a number of years as companies try to keep profitability high, but it's reasonable to expect numbers from 2% to 6% of the employee's salary. Be aware that the corporate 401(k) plan also has a vesting period for these matching funds.

- The number of companies offering a corporate pension has also been declining for a number of years, but they still exist. Again, there is usually a vesting period. You must stay with the company for a number of years before you are eligible for a corporate pension.

- There can also be miscellaneous benefits such as a company-subsidized cafeteria and a free reserved parking space.

Several of the aforementioned benefits involve a vesting period. An employee who leaves the company too soon may lose all or part of the benefit. It is worth noting that the company can also leave you. The company may sell your division, and this may prevent you from continuing to vest. Through no fault of your own, you may lose a significant amount of money in stocks, pensions, or other benefits. It may be worth trying to negotiate an employment contract that precludes such loss prior to accepting a new job.

You can also have a number of deductions from your paycheck:

- Parking fees.
- Contributions to a coffee or soda fund.
- Medical, dental, vision, life, disability insurance, and more. While these are generally considered a benefit, the total amount of money deducted from a paycheck can be staggering.

Remote Development

Working remotely, perhaps from home, can be a viable alternative to relocation. It is well demonstrated that remote workers can be very productive and successful. Cynical observers say this is because the boss has a much harder time distracting the remote worker with direction changes and status meetings. The productive remote worker calls into the meetings and continues working through much of the meeting.

Cynicism aside, remote development automatically results in more written direction and clearer specification of requirements and goals. This dramatically improves the chance of success of any employee.

Noncompete and Invention Agreements

Some companies require employees to sign documents stating they will not go to work for a competitor or compete against the company for several years. In years past, these agreements had limited enforceability because state and federal law generally recognized that individuals had to work to continue a reasonable existence. Some companies added clauses requiring employees to agree that enforcement of the agreement would not materially affect their lifestyle. For several years a number of companies have generated increasingly draconian employment agreements.

It can be a good idea to request copies of all employment agreements for review prior to accepting a new position. Feel free to mark up any desired changes. The company can always say no, and you can always work somewhere else.

Show Integrity

The best way to ruin job negotiations is to continually change your demands or add new requirements after the employer makes a suitable offer. You should be sincere in your statements and commitments. Showing bad faith in negotiations may taint your reputation and career for many years and many jobs. It is remarkable how many times bosses migrate to other companies for whom you may want to work. In a competitive market it is difficult enough to convince a prospective employer that you have the qualifications for the job, much less to overcome a reputation for lacking sincerity or integrity.

II

Product Development

Some men see things as they are and ask, "Why?" I dream things that never were and ask, "Why not?"

Robert F. Kennedy

6

Product Development Overview

Introduction

Product development is one of the more interesting things engineers do. The ability to bring something new into the world is very exciting, and it is an ability of which to be genuinely proud. Few engineers worthy of the title are unmoved by the delivery of their "baby" to the customer. This baby may be a communications satellite, a television set, a microprocessor, or a website. No matter what it is, what makes this so wonderful and exciting is that the baby is something you helped make. The feeling of being a proud parent of some new product is one of the great pleasures in engineering. Possessing the skill to do this well is a gift to be treasured.

So, how does all this product development stuff work? It turns out that doing product development for a corporation is dramatically different from what you might visualize in high school while thinking about becoming an engineer. Do a bunch of engineers sit around and build things? Ah, wouldn't that be nice. That might actually happen if a couple of friends are working in someone's basement. In a real corporate environment, however, there is much more to do and way, way more overhead and politics. The process of developing a new product within a large corporation is far more complex and has far more overhead than might be assumed. Launching a new development effort in a corporate environment involves meetings and presentations. Important people must be convinced of the merits of the idea, a consensus must be built, project plans must be created, and resources must be assigned. There must be regular status meetings, presentations, and lots of coordination with other groups such as marketing, sales, and manufacturing. All this takes time and often involves a fair amount of posturing and politics. The entire process can be quite intimidating for a socially unskilled engineer.

This part of the book takes a close look at the development of a technical product in a corporate environment, examines the most fundamental components of developing such a product, and discusses some of the overhead typically associated with development of a product in a corporate setting. As part of this we consider things that go wrong and why. Because this book is about engineering career management rather than product management, these discussions focus on ways for junior engineers to avoid obvious

mistakes and to enhance their careers while they are part of a product development team. Some of these junior engineers may someday become senior executives, so a few strategies are included to help them avoid being an impediment to efficient technical product development when they reach the executive ranks.

Social Interaction

People who never worked in an engineering environment may have an idealized view of how new products are created. They may think of groups of people wearing white lab coats and carrying clipboards, or they may think of a guy with long hair drawing a schematic on a coffee-stained paper napkin. Either of these can actually happen, but most often creating a new product within a corporation involves building a consensus among coworkers and superiors and getting approval of financial people to spend the money. What this really means is that social interaction is involved. Oops. This could be something of a problem, because many of the high school students who went into engineering were not known for their social skills. The engineering curriculums mastered by these students generally focused on needed math and science. We have, in effect, a selection and training process that ignores the importance of social interaction in the engineering workplace. This process often results in intelligent and talented people arriving in an unexpected and even undesired environment. Many engineering graduates by basic personality—and reinforced by many years of formal schooling—encounter the corporate world of product development poorly prepared to succeed in that environment.

The corporate world of product development sometimes involves much more than mere social interaction. Sometimes you must become a product champion and sell your idea and sometimes you must deal with a competing idea from a different person or group. Sometimes the competing idea comes from your boss or perhaps your boss's boss, and convincing your superiors that you have a better idea involves very delicate social and political interaction. *Selling* and *social interaction* are not concepts often associated with engineers, and herein lies the root of a problem.

Most people don't even dream of the salaries that are paid to young engineers. From the perspective of the vast majority of our civilization, young engineers are very successful people. There is every reason to believe that engineers have achieved this level of success because they can solve differential equations and because they survived courses on electromagnetic wave theory and other exotic mathematical and scientific topics. Unfortunately, however, nothing in their formal education has prepared them for the corporate political landscape. The needed sociopolitical skills are not taught in school. Rather, they can be learned playing video games with friends, dating, and hanging out at the pool during summer vacation. As smart as they are, some percentage of engineers are lacking in these social skills and, worse, don't appreciate the need for them.

People become engineers by surviving a challenging technical curriculum in college. In general, they were accepted into this challenging college curriculum because they showed skill in math and science in childhood. Powerful selective forces were at work their entire life. Much of the material mastered by engineers involved the ability to focus on and to solve problems in solitude. Solving an engineering problem often involved sitting and thinking and flipping back and forth in the book trying to understand what was going on. The social interaction of having someone tell you how to solve the problem did not always help to solve the next problem. You had to figure it out and understand. Contrast that, for example, with creating music or a play where much of the learning involved presenting your creation to an audience and understanding their reaction and learning how to manipulate or create a desired reaction. The educational process that produces an engineer is a formidable selection process for personalities and instincts to work with facts and solve problems with a step-by-step logical process. In the workplace this is very different from the sales, marketing, and business personalities who see negotiation and "doing the deal" as paramount. Even within the close confines of the engineering group itself, engineers with social skills are able to establish a bond with executives and manipulate their less adept coworkers. Great creativity and technical talent are sometimes not as certain a path to a promotion as inviting the boss over for dinner or coaching his child's soccer team.

When the time comes for management to choose an approach for a new product, it makes the very reasonable decision to choose one advocated by someone it trusts. Contrary to the image portrayed in Hollywood movies, the brilliant maverick rarely wins this popularity contest. While management may respect the technical skill of "that wacky guy on the third floor," it doesn't really trust him. Furthermore, the marketing and sales groups are viewed as having a closer connection and better understanding of the needs of the customer.

Finally, management has learned through experience that approaches proposed by engineers may sacrifice desirable features for the sake of an easier implementation. Those in power act on product ideas proposed or championed by those who are trusted. The sad reality is that a mediocre or even bad idea proposed by a trusted individual will be chosen over a wonderful idea proposed by an engineer not held in high esteem by the corporate management. This is not something malicious but quite practical. Senior folks in the company, even if they once were competent technically, have been too busy in corporate management to stay abreast of the newer technology. They just don't know if something is risky or easy with the newfangled technology. They must trust the opinions of others, and most often it is social adeptness and not technical competency that engenders such trust. Let me make sure that the message is clear. This is not the fault of the corporate management and is certainly not malicious. It really has no other rational option. By basic temperament and an educational selection process, engineers as a group are simply not the most gregarious of people. The result is they often don't take an interest in forming the social bonds needed to positively influence senior corporate management.

Product Development Blueprint

Social politics of the real world aside, how should a new product be developed? One might envision a blueprint for the successful creation of a technical product. That blueprint might resemble the following sequence of events:

1. **Idea:** Get an idea for a product. This may be "blue sky" daydreaming or may be in response to a perceived customer need.

2. **Money:** Figure out if it will or how to make money selling it. Under some circumstances it may make sense to invest in building a product that doesn't make money. In general, however, assigning personnel and spending capital to create a product benefits a corporation only if there is a rational belief that it will bring a worthwhile amount of money into the corporation. A "worthwhile amount of money" varies for corporations of different sizes, for product lines, for the proportion of the workforce currently engaged, and for many other reasons.

3. **Strategy:** Once it is determined that a product might generate a worthwhile amount of money, choose a high-level but comprehensive development and marketing strategy. This strategy should include an evaluation of the time to market, a preliminary cost to develop the product, risks, the demographics to which the product will appeal, a basic marketing approach, and an estimation of the cost of acquiring a customer.

4. **Requirements:** If, after factoring in anticipated development and marketing costs, it still looks like the product can generate an interesting amount of revenue, the product idea should be refined, a product requirements document should be written, and a more detailed analysis should be given.

5. **Plan:** With the product specification (the requirements document) in hand, a detailed engineering analysis can be done; the necessary staffing, development schedule, and budget can be determined; and a detailed project plan can be written. Reasonably accurate product budgets and schedules finally allow a comprehensive management review of the costs and pros and cons of the product idea. If everything looks good, it is time to actually start the project.

6. **Begin:** Assemble a multidisciplinary team to ensure that all tasks (from writing a user manual to ordering components and shipping boxes) are executed by knowledgeable people. Make sure the team is reasonably well focused on the new task and not distracted by legacy projects.

7. **Design:** Produce a technical design document. Have appropriate experts review the document, and make sure the design accommodates the product requirements.

8. **Manage:** Manage the product development effort to ensure that tasks are executed in a coordinated fashion. A critical component of successful management is accurate reporting of statuses by all organizations. This allows the project manager to adapt to changing realities and to realign resources to avoid unnecessary waste.

As astounding as it may be, I've never seen a project run this way. Some of the projects on which I've worked have failed nearly every one of these steps. How can this be? You are dealing with highly intelligent, highly educated, and highly motivated people. How can capable people make such obvious mistakes? The answer is amazingly simple yet amazingly complex. The answer once again involves social interaction instead of technology. Projects can fail for genuinely technical reasons. However, my personal experience is that true technical failure is rare and that risk can usually be contained. Failure most often results from mistakes, excessive optimism, oversights, stubbornness, pride, and other basic attributes of humans. Such problems are much more difficult to predict and are offset because the notoriously capricious human nature is involved.

7

Basics First

Introduction

With most human activities good things happen when the fundamentals are addressed early. Product development is no different. A large mass of data proves that projects are significantly more successful when early work clarifies the big picture of what needs to be done and specifies a reasonably detailed plan for doing it. Establish the big picture, lock down the basics, and you are well on your way to a successful project.

This chapter expands on the first four stages (idea, money, strategy, and requirements) of the product development blueprint introduced in Chapter 6. These are the foundation of a successful product development effort.

Idea

Product ideas arise in every discipline and at every level of a company. But, with a tip of the hat to *Animal Farm*, some ideas are more equal than others. A proposal for a new product rarely stands on pure merit because many of the people involved in the review process don't have time or perhaps don't have the background to fully understand details of the idea. In some situations engineers are perhaps the most qualified to come up with a technical product idea but may lack the social skills to convince the company of its merit. While engineers were studying calculus, the people for whom the engineers will be working were out partying and honing their social skills. Often positioning and connections are more important than the actual quality of the suggestion. Initiatives that originate with senior executives carry far more weight than an idea from a summer intern. A senior position and a persuasive or forceful personality can go a long way toward furthering the cause of a proposed product. Forceful, gregarious, and persuasive are not often viewed as hallmarks of the engineering personality. To be effective in the corporate environment, engineers must be able to build a rapport with

peers and superiors. Engineers would do well to improve on their ability to write and speak persuasively and to interact well and build social alliances.

At this point a caveat is in order. It may not be career enhancing to put forth a persuasive argument that demonstrates your proposal to be better than that of a superior. You would like to believe that good suggestions are recognized for their value and that articulate and competent engineers are praised. Unfortunately, this is not always the case. Some personalities view having another initiative chosen over theirs as public embarrassment. Others perceive the victor as a rival rather than a resource. There are also a variety of generally insecure personalities who do not respond well to having an underling "win." Competing openly with a superior can be a bad idea for many reasons. A popular solution to this problem is to quietly go along with whatever the boss says. Although this approach allows engineers to avoid conflict with superiors, it has unfortunate consequences for the engineers and the corporation. Talented but timid engineers who just go with the flow remain obscure and therefore unappreciated members of the corporate structure. Although the engineers avoided conflict they also did not distinguish themselves from the other employees of the company. In addition, by not speaking up, the engineers may allow the company to expend resources in developing a suboptimal product.

It is never a good idea to publicly challenge a superior. Doing so generates an assortment of career damaging messages. Furthermore, being technically correct means little in the political landscape of a large company. Engineers must accept that there are very few corporate situations where they make the final decisions. Public challenges or confrontations are never a good idea. Instead, privately approach the executive with your proposal. Depending on the personality of the person you are appealing to, it may even be a good idea to take a submissive or subservient posture. In extreme cases there is simply no good way to broach a counterproposal to the demigod in charge. Be aware of the personalities of those around you, and never assume that being correct will make your idea welcome. In general, executives do not like to be told they are wrong, and they certainly do not like to be lectured. If this seems like a lot of politics, you are right. Engineers should save the math and science for making the project work. The corporation is made of people who interact socially and politically with pack leaders, dominance displays, and personal vendettas. Behind-the-scenes deals with power brokers can enormously benefit the engineer's career. Socially adept engineers can get the company to use their ideas and to view them as valued members of the team. Politics and social hierarchies are an intrinsic part of every human culture, and they are certainly part of every corporation. The pack leaders and the dominant personalities seize control wherever they go, and they really want people to do what they are told. A corporation is not a democracy, and the meek have not yet inherited the earth. Engineers must accept this and respect their place in the corporate hierarchy or risk serious career damage.

Money

It takes money—sometimes a significant amount—to develop a new product. To make this investment worthwhile, a company must have some reasonable assurance that it will be able sell enough of the product at a high enough price to recover the initial investment and to make a profit. Sometimes corporate decision makers fall into the belief that there is a very small market window, and the product must therefore be developed too quickly to spend time analyzing its profit potential. Sometimes the idea behind the product is so obviously good that an analysis of the profitability prospects is omitted. Unfortunately, a number of these obviously good ideas end up being disastrously unprofitable, and those associated with such corporate disasters suffer damage to their careers. Every corporation of any longevity has had its share of such failures. However, corporations that excessively penalize failure create analysis paralysis where executives are afraid to take a risk lest they become lepers with stigmatized careers. Most executives, indeed most people, simply choose a low-risk path even if it means a mundane product or existence. All things considered, it can be quite difficult to determine whether profitability analysis is worth the cost and delay. A variety of unpleasant corporate situations may exist with respect to the profitability of a proposed new product:

- The product idea originated with senior managers, and few if any want to incur their wrath by questioning the viability of the product.
- The product advocates exert influence to ensure there is no negative image of their product concept.
- The profitability analysis was done, revealing that the product would likely make money, but not enough to be the correct application of resources for the size of the company.
- Some component of the profitability analysis depended on another group in the company, and that group was being unresponsive.
- Profitability ultimately hinges on the magical development or deployment of a nonexistent or high-risk technology.

Making a profit with the product generally involves some form of sales and marketing plan. The corporate environment tends to isolate marketing and engineering activities so the typical large company engineer may not often be exposed to detailed product marketing plans. This makes sense from the perspective that engineers should be busy doing engineering and should not have time to participate in positioning the product in the marketplace. Likewise, the marketing folks should be busy and not have time to help design the electronic circuitry. Here again, however, we may have social interaction and defensive posturing that engineers are not expecting. In general, it is obvious that the marketing folks cannot design the hardware or software of a technical product.

It is not so clear that engineers cannot help with the marketing. In fact, some engineers have an impressive understanding of the market for technical products. Often, they are the "early adopters" and buy or use many of the toys. Nevertheless, engineers must understand that making marketing suggestions, even valid ones, is tantamount to inciting a turf war. Some marketing and business folks are former engineers, but most cannot even pretend to do what the engineers do. While it is probably not true that engineers are smarter than marketing people, a marketing person may silently or subconsciously feel that way. Challenging favored business or marketing assumptions can make the marketing person feel threatened. Socially oblivious engineers who continue to challenge may end up in a career-damaging discussion where the marketing folks take offense and see a need to put the engineers in their place. In such situations the presumptuous engineers may be summarily dismissed as not understanding the business. At this point, continuing to question the accepted approach, even if the questions are valid, becomes really dangerous to career advancement. Later being proven correct can be even worse for the engineers as those questioned could become extremely defensive.

The hope is that an executive responsible for the well-being of the company would welcome a discussion of possible flaws in the current plan. Unfortunately, this is not always the case. In a large corporation, continued advancement is often associated with smooth, problem-free operation of your group. Management is paid to solve problems, not to propagate them to the next level of management, and certainly not to create them. Controversy, mistakes, or exposed poor planning may significantly disrupt future career goals. What many socially unconscious engineers fail to grasp is how important being correct is to some people. When this basic personality is magnified by career aspirations, innocent questions by engineers may yield a venomous response. It is possible that marketing has not done its homework, or perhaps the entire approach is based on wishful thinking. Engineers must understand that some people have enormous egos and would sooner destroy the company than admit they are wrong. A corporate executive with such a personality simply will not brook challenges and will never admit error. Take the hint. Don't argue. Walk away from the discussion with your career intact.

There are also situations where the engineer is not aware of some facts that resulted in adopting the current approach or may not be thinking about things the right way. Certainly taking everybody's time in a meeting to correct this mistaken viewpoint is inappropriate. However, an interested executive may schedule some time to privately discuss the subject with the engineer. Directly addressing engineering concerns improves morale and leads to a tighter integration of development effort with the corporate goals. Part of management's job is to mentor and lead, so simply dismissing the engineer's question is making a statement that the engineer is not worth the time needed to correct the mistaken viewpoint. Should this happen, the engineer must understand the meaning (he or she is not considered worthy) and must incorporate this reality into his or her behavior. If the engineer is unimportant to a manager, so be it. Become important to another manager, or employ some strategy to develop a personal relationship with the manager. Hint: Being clever and pointing out potential problems with established plans is the wrong approach for many executives. They really don't want to hear about problems with their ideas. They want you to do what you are told, and they want to be able to trust that you will never make them look bad. One educational aspect of my engineering career was to learn that the parody of the corporate yes man was not a parody but a reality. A number of corporate executives really do not want anybody to correct or debate them. Accept this. It is a reality, and you will not be able to change this. Focus on adjusting your behavior to become nonthreatening and supportive of their ideas. This is the only way you will be trusted by them—and therefore the only way to positively influence them.

A surprising number of projects get launched that have no genuine possibility of earning an interesting amount of revenue for the company. However, it is occasionally a good idea to proceed with the development of a new product even if profit appears unlikely:

- The image of the corporation requires a product for a particular market. This makes some sense so long as the need is not a fantasy and product development is not so expensive that it damages the financial stability of the company.
- The money-losing product is a high-end loss leader that establishes the premium device in the market and draws customers to more economical models in which sales can subsidize the incurred loss.
- Although development of the product costs more than the expected revenue it will keep the staff engaged and challenged during a slack period.
- Although development of the product costs more than the expected revenue it improves the company's position in recruiting top talent.
- The product will prevent a competitor from gaining market dominance.

A discussion of product profit potential would not be complete without some consideration of intellectual property rights and their protection. Innovation would be slowed or halted if a competitor could simply copy the complicated internals of a product and start producing it. To survive, the corporation must be able to sell sufficient product at a high enough price to recoup the cost of development. Numerous governments and ruling organizations realized this a long time ago and created a variety of intellectual property protection mechanisms including trademarks, patents, and copyrights. Specialists can spend a lifetime arguing the details of these things but a design engineer should have some understanding of what they are and their value.

Actual legal action against a company can be expensive and time consuming. Instead, corporations often use an extensive patent portfolio as a negotiation tool in achieving favorable terms when working with another company. Toward this end, many corporations encourage engineers to come up with potential patents by offering token pay for good ideas. Over years, and with relatively little expense, a formidable portfolio of intellectual property can be amassed that adds considerably to the value of the company.

The cost of some development truly requires legal protection of intellectual property to ensure the continued survival of a corporation. However, considerable advancement of the human condition has occurred because someone looked at an existing product or design and made it better. Indeed, part of the original idea behind a patent (as distinguished from a trade secret) was to make the information available for public inspection while offering market protection for the patent holder. Some current social attitudes and laws make risky the time-honored technique of inspecting an existing design and improving it. Indeed, engineers in our modern litigious society need to be aware of things like the Digital Millennium Copyright Act that essentially say the engineer is not allowed to examine the operation of an existing implementation. Some techniques (well beyond the scope of this book) allow the independent legal re-creation of existing technologies, but this is a legally complex field that needs to be treated carefully. In summary, protection of intellectual property rights is necessary for the survival of companies but can sometimes interfere with righteous and ethical innovation and may be detrimental to rapid advancement. It is reasonable to believe that the trade-off among protecting the revenue stream of corporations, ease of innovative development, and customer convenience will continue to evolve and to generate controversy for many years to come.

Strategy

Laying out an initial strategy is an exciting time in the life of a product. It is at this stage that we see the newborn idea take on a life of its own. What corporate resources are needed to deliver this product in a timely fashion? How soon can it hit the market, and what competitors are likely to exist at that time? Given the expected competition, what is a realistic sale price of the product? How much will it cost to develop and how much to manufacture? Can some features be added in a timely fashion to provide market differentiation? So many questions to answer. So much strategy to prepare.

A poor product strategy, or perhaps a poorly executed good strategy, can cost the corporation a great deal of time, effort, and money and can doom the product. A critical part of creating a product strategy is determining risks that may be encountered. For a technical product designed by engineers, this certainly requires an accurate assessment of technical risks and approaches that may be used to minimize these risks or to overcome them should they occur. Of particular importance to the engineers are technical risks that, should one be encountered, result in the irrecoverable failure of the development effort. Corporations occasionally begin development of a product with one or more serious risks. Sometimes this happens due to the ego of the executive who conceived the product, and sometimes there are legitimate corporate reasons. Engineers assigned to a high-risk development effort are actually in some danger themselves because career advancement can be adversely affected by association with a failed project. Even if the engineers' part of the project was delivered ahead of schedule and exceeded performance expectations, their image can be tainted by the overall negative aura of the failure. From a career perspective, it is sometimes better to be a mediocre engineer on a successful project than a great engineer who was unable to rescue an ill-fated one.

Socially adept engineers (or perhaps those who have simply been burned by previous association with a project disaster) may try to avoid high-risk projects. Such resistance creates a problem for the project and necessitates a staffing strategy. Ordering resistant people to work on a favored but risky project is generally not good for morale and therefore doesn't readily lead to a productive development environment. Corporate management often approaches this resistance with a strategy of minimizing the level of risk assigned to an effort. Talented engineers, however, can readily determine the true nature of the technical risk. After all, it is their job to understand nature and to determine ways to bend it to their will. They must work in the world of real facts and real science.

It doesn't take long for a knowledgeable engineer to determine that some risks are being falsely minimized. The developers with the deepest knowledge of the most serious problems know exactly where the project stands, and it doesn't take long for such a person to write off a cheerleading executive as a clueless or posturing bureaucrat. This is unfortunate, because these are exactly the talented people on which the future success of the project most depends. The good news for the executive is that most engineers are nonconfrontational and will simply go about the business of trying to rescue the project in silence. The project staffing strategy of publicly minimizing risk is, therefore, generally successful. In a rare instance, one may encounter an engineer with a forceful personality who also lacks the social awareness (some may say common sense) that publicly challenging a superior is a bad idea. Corporate management gets paid to see that its agenda for corporate success is carried out. In this light, it is entirely reasonable that management views the challenging engineer as disruptive, an obstruction, and not part of the team. That the engineer is technically correct is irrelevant. He or she is interfering with the vision put in place by corporate management. This is most easily addressed by no longer inviting this person to meetings or labeling the person as hard to work with or as a malcontent. Career advice for the engineer is this: If management cares what you think about project risks, they will ask you; otherwise just do your job.

It is not just engineers who may shy away from high-risk projects. Those who must authorize funding for the project may be disinclined to do so if the perceived risk is too great. Vendors who supply components used by a risky project may hesitate to commit to large volumes or discount prices. Corporate management will minimize the true risk of a favored project for many reasons, and will use many strategies for doing so. These strategies should be viewed as risk cosmetology and should be distinguished from risk management, which is more of a true engineering exercise. Risk cosmetology is a widespread practice. It keeps shareholders, financial analysts, and others at bay while the engineers try to solve the real problems.

The inherent technical risk of a product development project can be compounded by placing the project on a compressed schedule. A short or aggressive schedule compounds risk for many reasons:

- Shortcuts are taken, and process steps are skipped.
- Known risks are not thoroughly investigated and resolved before completing large amounts of well-intentioned but misguided work. Some of this completed work may have to be thrown away to fix underlying problems in later stages of the project.
- Haste may cause early symptoms of problems to be missed that would have been noticed in a more methodical atmosphere.
- Engineering panic—that is, engineers operating under stress and tight deadlines are more susceptible to making mistakes.
- Calculated risks are taken. In this situation, a rapid implementation that might work is used instead of a proven solution that takes longer to implement. The worst-case scenario here is that the rapid implementation mostly works and that resulting problems are difficult to reproduce and correct.
- Observed quirks and anomalies are ignored in the drive toward getting something that can be demonstrated. Some of those may become serious problems that take much longer to resolve once overlaid with additional complexity.

It seems to be a bad project strategy to force an accelerated schedule. Why do this? Several possibilities follow:

- Engineers are expensive. Nobody wants them goofing off.
- The company wants to minimize the development expense and start making money (selling the product) as soon as possible.
- The management philosophy, "The perfect airplane never leaves the hanger"—that is, if you don't give the engineers a deadline, they will keep tweaking and improving the product forever.

- Any lucrative market always has competition. Getting to the market before other companies can bring a greater share of the market and presumably greater profit.
- A specific marketing deadline such as the end of the government year or the Christmas buying season.
- A physical deadline such as, "The planets will not correctly align for another 11 years and the planetary probe is launching on the appointed date—with or without your device."

Some—perhaps all—of these are good reasons to set an aggressive development schedule. Striking the right balance between aggressive and methodical development is one of the things corporate managers get paid to do. If they do a good job their company will produce innovative quality products and have high morale and low turnover in the engineering staff.

Determining and managing product development risk is important, but also important is achieving product profitability commensurate with the risk. That is, a risky but successful development effort should produce a highly profitable product. Early product strategizing is most beneficial if it can determine reasonable estimates of the likely profitability of the product. A simple-minded way to look at profitability is the sales volume of the product times the sale price of the product less the various continuing costs of making the product less the nonrecurring initial expense of development.

Costs of Developing the Product

Developing a product is generally not viewed as a continuous effort. It has specific phases. The work flow moves from design to implementation to testing to manufacturing and then into shipping and warranty repair. Development activities include the following:

- Industrial design engineering—that is, designing the attractive enclosure of the product
- Hardware design engineering, including selecting and qualifying components, designing the circuitry, ensuring the box doesn't overheat, and overseeing agency certification (e.g., Federal Communications Commission, Underwriters Laboratories)
- Software design engineering, including software that manipulates the hardware, sends and receives data, controls the system, and the implementation of the user interface
- System integration and testing, where the various subsystems are pulled together and tested as a complete system
- Project management, which is responsible for setting priorities, maintaining an adequate staff, resolving conflicts, and tracking status
- Writing a user manual, designing packaging, and shipping material

- Creating manufacturing test fixtures, test plans, and test automation and establishing a plan to feed the flow of components into the manufacturing process

Costs of Manufacturing the Product

The recurring cost of making and selling the product includes the following:

- The cost of the components and raw materials used in the product. The more that are made, the more raw materials are needed. However, there are often economies of scale such that the cost of each product becomes progressively cheaper as more are built.
- The labor and equipment involved in manufacturing the product. Once again economies of scale come into play. Greater product volume justifies more sophisticated and complete factory automation.
- The cost of royalties paid to license intellectual property used in the product. Intellectual property (i.e., patents) can be licensed in any number of ways including per unit (each product costs the same), tiered (as more units are sold the cost per unit goes down), and a one-time buy-out where a flat rate is paid no matter how many products are sold.
- Cardboard boxes, packing material, user manuals, installation CDs, and so forth.
- Warehouse space for storage of components used to build and package the product and also for storage of complete products waiting to be shipped.
- Shipping costs of delivering the products to customers.
- Sales training, sales commissions, and so forth.
- Web site for information, ordering, etc.
- Billing systems.
- Advertising.
- Customer service and customer service training.
- Warranty repair and analyzing product returns for failure patterns.
- Maintenance engineering to identify fixes for failure patterns and to replace components that become obsolete.

Special Note on Component Pricing and Schedules

If you ignore the issue of royalties, software reproduces very cheaply but then you have the continuing expense of buying mechanical and electronic components used in making the product. In many cases, the cost of a component today is higher than the cost of that same component a year from now, when the product design has been completed and the units are in mass production. For this and other reasons, the cost of the product at the start of production when it is first introduced may be higher than when the product has been successful and is being sold in high volume. There is science, art, and more than a little magic in estimating these future prices in a technique called "forward pricing." There may also be deception.

The recurring cost of software used in a product (other than royalties) is very small. Forward pricing is therefore mostly a hardware activity. Sometimes the hardware group has a strongly preferred design approach. To ensure that its preferred approach is used, the forward pricing of components is artificially manipulated to indicate that its preferred approach is the cheapest. Corporate management should keep an especially watchful eye for forward pricing deception in the case where the hardware and software groups disagree on the correct implementation of the product. Forward pricing manipulation can also be effective in blocking or perhaps instigating specific product features or marketing initiatives. Schedule estimates can also be intentionally skewed to encourage or discourage product features and architectures. Unlike forward pricing, essentially any group involved in the project can artificially manipulate schedules to encourage its specific agenda.

Detecting manipulation of pricing or schedules can be very difficult, perhaps impossible, without specific knowledge of the technologies involved. The unfortunate reality is that these manipulative strategies work quite well, as many executives overestimate their knowledge of the underlying technologies and overly trust those managers who deliver the estimates. Some executives never visit the cubicles and laboratories and never talk directly to the engineers involved in a project. Doing so can be very enlightening and in many projects would contradict some portion of the schedules and status the executives are receiving. Executives' inherent trust in the management hierarchy means, in effect, that scheming engineering managers hold the keys to the feature set of every project they oversee. In the hands of a visionary this power can bring amazing innovation to the market. A more conservative, risk-averse individual can force the company down the path of bland copies of successful products already on the market. When carried to the extreme an entire innovative project can be killed by providing pricing or schedules incompatible with the corporate goals.

There is an even darker side to schedule manipulation. In situations where the senior folks provide loose oversight, schedules can be used to embarrass opponents or to enhance the status of favored employees. A powerful manager insisting on a short schedule can damage the reputation of a hardworking and competent engineer who is unable to meet the contrived schedule. It is unlikely that the besieged engineer will even be at the meetings where the schedule and his or her inability to deliver are presented and discussed. Conversely, favored engineers can be given liberal schedules and may be invited to high-level meetings to present the good news of exceeding expectations and delivering ahead of schedule. For those engineers or engineering managers who are, shall we say, ethically challenged, schedule manipulation can be an effective strategy for personal advancement and for the advancement of those they support.

Estimation Responsibility Matrix

Someone has to create the estimates for the various ingredients that go into the development of a new product. Table 7.1 is a summary of the various responsibilities associated with providing estimates for both development and continuing production of an arbitrary example product. The "Responsibilities" heading is broken into four groups: hardware, software, program management, and other. Each of these groups is further divided into estimates for development and production. These eight categories encompass the responsibilities for estimating the cost of making a modern electronic product.

A notation in a column indicates that it is righteous and proper for that group to estimate the issue represented by the row under consideration. For example, estimation of the majority of the production-related issues falls under the domain of the hardware group. Software reproduces readily and very economically. In most corporate cultures, this disqualifies the software group from heavy involvement in production. As such, Table 7.1 shows an empty column for software-related production issues. Similarly, program managers' primary responsibility lies elsewhere, but they can realistically be involved in a variety of development estimates by setting expectations for the desired commitment and quality. It would be an unusual company where program management would be involved in production estimates.

A closer look at Table 7.1 reveals that there are actually two notations used in marking the columns. An X indicates that the category of items being considered has fairly standard pricing and is therefore not readily susceptible to the forward-pricing shenanigans described in the previous section. An O indicates that the feature is conducive to positive or negative forward-pricing manipulation to encourage adoption of a specific architecture. Obviously, details of a particular effort may convert X's to O's and vice versa. The point of this notational distinction in the table is to demonstrate that the engineering teams have significant clandestine ability to influence product features, approaches, and architectures.

Licensing and royalty considerations are something of a wildcard and are not represented in this table. These fees can be associated with hardware components, software modules, user interface concepts, and more. The decision to incur such fees is usually that of the business group, based on a recommendation by the development team.

Cost estimation issues and features are presented by rows in Table 7.1. The grouping of the elements is for illustrative purposes and is somewhat arbitrary. For example, "Custom FPGA/ASIC" may be included in the "Exotic Chips" category and "Printed Circuit Board, Connectors, Cables, Screws" could be combined with "Enclosure and Fans."

To be clear, it would be unusual in a company of reasonable size for actual engineers to negotiate the price of components. A good-sized company has a staff that handles this. However, the engineers specify the components and details associated with the components and in doing so establish the ballpark for the negotiations.

TABLE 7.1

Estimation Responsibility Matrix

Issue	Responsibility							
	HD	HP	SD	SP	PD	PP	OD	OP
Hardware Development (Selection and Interconnection of Circuits)	O							
Software Development (e.g., Operating System, Drivers, Application)			O					
User Interface Development			O				X	
System Integration and Testing	O		O		O		X	
Passive Components (e.g., Resistors, Capacitors)	X	X						
Memory Components	X	O						
Processors	X	O	X					
Exotic Chips	O	O	O					
Custom FPGA/ASIC	O	O	O					
External Interfaces (e.g., Ethernet, USB, RS-232)	X	X	X					
Printed Circuit Board, Connectors, Cables, Screws	X	X						
Power Supply	X	X						
Enclosure, Cooling, Fans	X	X					X	X
User Manual			X		X		X	X
Formal Release Testing	X		X		X		X	
Manufacturing, Test Fixtures, Final Testing	X	X	X		X		X	X
Packaging and Shipping Material	X	X			X		X	X

Note: X = the category of items being considered has fairly standard pricing and is therefore not readily susceptible to the forward-pricing shenanigans described in the previous section; O = that the feature is conducive to positive or negative forward-pricing manipulation to encourage adoption of a specific architecture; HD: Hardware team's estimated cost to develop; HP: Hardware team's estimated cost to produce; SD: Software team's estimated cost to develop; SP: Software team's estimated cost to produce; PD: Program management's estimated cost to develop; PP: Program management's estimated cost to produce; OD: Other group's estimated cost to develop; OP: Other group's estimated cost to produce

Hardware Development

This includes selection of hardware components and creating a hardware design that meets the environmental, functional, and operational requirements of the product under development. Implementation of the design to the point of system integration is also included. This category is that of nonrecurrent engineering (NRE) and does not include ongoing support of the product during the production phase. Only the HD responsibility column is marked.

Software Development

This issue is the software NRE complement to the hardware effort. It includes selection, design, and development initiatives necessary to meet the requirements of the product. Only the SD responsibility column is marked.

User Interface Development

User Interface (UI) development may consist of software development working in conjunction with graphic artists, human behavior specialists, focus groups, and other experts. Responsibilities here include software development (SD) and research and development contributions from the other specialists and experts (OD). As such, the SD and OD have been marked as responsibility columns.

System Integration and Testing

System integration is where the product as a whole comes together and everyone gets into the act. SD and hardware (HD) developers work out their misunderstandings and smooth their interface. Other groups (OD) may be involved in resolving problems or arbitrating disputes. Program management (PD) may weigh in to prioritize bugs, to waive requirements, and to emphasize the importance of features and goals. All of the development columns but none of the production columns are marked.

Passive Components

This is the first issue with impact on the estimated production cost of the product. Extensive development and system integration activities have no intrinsic impact on the production cost, but every resistor and capacitor increases the eventual product cost. The HD and HP columns are marked because estimation of the cost of the passive components to build prototypes and later volume production falls into the domain of the hardware team. Passive components, however, have fairly standard pricing and therefore are not typically used to encourage adoption of a specific architecture.

Memory Components

New memory technology is introduced on a regular basis. Some of the new technologies are successful and dominate the market until the next innovation takes over. Estimating future cost of memory components requires knowledge of current technology trends and mystical anticipation of future trends. Estimates for components of this type can be invisibly adjusted to encourage architectures and approaches favored by those providing the estimate.

Processors

Like memory components, new versions of processors with varying features appear with frightening regularity. Unlike memory components, software folks also get into the estimation fray because they must offer an opinion on the difficulty of creating and debugging software on a specific version.

Exotic Chips

Exotic chips can consist of a number of devices, including custom field programmable gate array (FPGA) and application-specific integrated circuit (ASIC). These components offer all the bad estimation features of memory components and processors. Then it gets worse. Newly introduced or custom made chips offer a wealth of opportunity to reduce cost and save time. Some will live up to the expectations and more. Others will catastrophically fail. It's hard for experts to predict which will occur, and nearly impossible for the layperson.

The remainder of the table consists of relatively mundane components and commodities such as printed circuit boards and packing material. A number of these issues involve collaboration with other groups such as technical writing, testing, and manufacturing. These commodities generally have predictable pricing and are not especially conducive to manipulating costs to encourage a favored architecture.

Special Note on Cost-Reduction Activities

The pressure to reduce the production cost of a product is relentless. Sometimes this pressure is relaxed somewhat for the sake of expedient delivery of the first version of the product. Once delivered, a variety of cost-reduction proposals are both demanded by corporate leaders and submitted by engineering managers. Product cost reduction is an opportunity to save the company money and, therefore, an opportunity for the manager of an engineering group to shine. A group awarded the cost-reduction effort may need more staff to accomplish the additional work. This is a career opportunity for the manager of the group. Higher head count means greater responsibility and at some point a promotion. Thus, the company benefits from a lower-cost product, and the manager gains respect, responsibility, and is rewarded for having helped the company. This is a win–win situation. Or is it?

Not all cost-reduction efforts benefit the product or even the company. Just because a product is cheaper to produce doesn't mean the company will make more money. One must be careful of the nature of the cost-reduction strategy, its consequences to the customer acceptance of the product, and the expense of the cost reduction effort itself. Some examples are as follows:

- The manager of the group that makes custom integrated circuits for the company proposes making a new ASIC that performs the job of several components used in the current design. The cost of the proposed ASIC is $6 cheaper than the cost of the replaced components. Because the company expects to sell 200,000 of these products next year this appears to be a savings of $1.2 million. However, developing the ASIC will cost $1 million. Does this mean the company will save $200,000 by making the ASIC? Perhaps, but forward pricing of the existing components must be considered. What the components cost this year is not so important as what they cost when the ASIC is eventually ready. There are also costs associated with incorporating the ASIC in the product. A new hardware design must be done, and a new circuit board must be fabricated. Because components moved around on the board the manufacturing fixtures must be changed. New software may be needed, and all this new stuff has to be tested. The new ASIC may end up not saving anything. However, developing it may still be good for the company if the technology used in the new ASIC is an important technology for the future of the company. These decisions are really complicated.

- A high-volume product made by a company has multiple buttons to allow users to navigate menus. An idea surfaces to remove some of the buttons to save money. Cost analysis indicates that removing the buttons will save $0.30 per unit. Because the company sells more than a million units per year, it will save more than $300,000 by removing the buttons. Further analysis indicates that software modifications to create a new "reduced button" user interface will cost $200,000, and other associated changes and testing cost an additional $50,000. Ultimately, the company will genuinely save $50,000 a year by removing the buttons. Unfortunately, removing the buttons, even with the software changes, seriously impairs the user interface in the eyes of the consumer. Sales plummet as customers purchase competing products from other companies.

- A consultant to the marketing group argues vehemently for additional product features to distinguish the product in the marketplace. The powerful hardware manager on the project continually rejects all but the most rudimentary features as costing too much to implement. After months of arguing the consultant quits and, as he is leaving, exclaims, "The cheapest product we can make is an empty box—but don't expect to sell many of them."

The point of these examples is that appearances can be deceiving and that genuine engineering trade-offs must often be combined with seemingly mystical insight into the future. Cost-reduction efforts are important, but some things are more important to the success and profitability of a product than just making it cheaply. Striking the correct balance between building a cheap or an exotic product is the job of corporate management, whose skill of getting the right advice and making the right decisions will determine the product's success in the marketplace and ultimately the success of the entire company. Some legendary corporate leaders have excelled at creating markets for new products whereas other companies are profitable simply making economical copies of existing products. The path to success is varied and bumpy. Anybody can get lucky, but companies that consistently produce successful products are well led. Putting your faith entirely in low prices and short schedules is limiting your options. Somebody can always make it quicker or cheaper. Few care enough to make it better.

More than Engineering

Creating a comprehensive product strategy clearly involves much more than engineering and manufacturing. It involves multiple divisions of the corporation and many aspects of marketplace and the demographics of the customer base. It involves considering the early adopters and distinguishing between the early adopters and mainstream customers. It involves an analysis of the competition, the resources of the competition, and the leverage the competition can apply to the marketplace. Finally, it involves corporate expenses, revenue streams, cash flow, and other purviews of the business group. However, this is not a book about corporations but about engineers and their careers. From the perspective of this book the most important thing to emphasize about the many groups in a corporation is the need to maintain an active and mutually beneficial relationship. It is important to create a feedback loop of engineering capabilities, marketing demand, and business goals and constraints. In such a relationship the engineers, marketing, sales, and business groups can work together refining the value of features and the difficulty, cost, and capability of current and near-future technologies. It is here that strategizing about the product helps define a comprehensive set of requirements valued by all facets of the corporation.

Requirements

Recall that, in general, corporations subscribe to the philosophy that the marketing and sales groups are more closely aligned with the needs of the customer than the engineers could be. This viewpoint certainly has some merit. The marketing and sales groups are in the field talking to their customers and are well positioned to understand what the customers want. However, an old adage in consulting is: success is achieved by giving your customers what they need, not what they ask for. The basic problem with customer-driven requirements (and therefore marketing- and sales-driven requirements) is that customers can reasonably ask only for things that they have already seen and understand. Customers are notoriously bad at visualizing the future and asking for something that doesn't yet exist but can be readily developed. Certainly any customers can say they want an antigravity device. However, only those familiar with current state-of-the-art technology can understand which innovations are right around the corner. Many of the great advances in consumer technology (e.g., the personal computer, the World Wide Web, the MP3 player) came about not because of customer demand but because some talented engineers envisioned the future. It may also be worth noting that countless such world-shaking innovations came from people who primarily wanted to do something they found "cool." Making money was not the primary motivation, and often the innovators made more than they ever dreamed possible. This contrasts sharply with normal product development in a large corporation where making money is of paramount importance.

In the profit-driven corporate environment technical innovations are more evolutionary than revolutionary. Major technological innovations produced by corporations really tend to occur because a senior executive, often the president, has a vision of the future and ignores the recommendations of financial advisors. Only the senior executives have the corporate power to proceed with "ill-advised" products. Famous examples of this include Akio Morita and the Walkman; Bill Gates, unhappy with cash cow DOS, betting his company on Windows; and Howard Hughes repeatedly risking corporate bankruptcy in a never-ending quest for greater aircraft performance.

While the wealth and legacy of risk-taking corporate leaders are renowned, most corporate undertakings are much less colorful. Although exceptions exist, processes at well-run corporations generally demand a requirements document be created to define the features and operation of a new product. A requirements document is not a design document. It says what a product is to do, not how it does it. In an ideal environment, a requirements document provides the ultimate reference for the product implementers. It describes the look, feel, and behavior under various circumstances and may even set performance standards. In a righteous world, the people who ultimately set the requirements (generally the marketing department) should write the requirements document. It has been my experience that this rarely happens. For various reasons, the job of writing the requirements document often falls on the engineering group. Engineers who write the requirements document bring, through basic personality and years of training, the precise mental processes to specifically and explicitly state what something does. Their connect-the-dots thinking process allows them to specify an exact definition of the product. Engineers who can also write well gain higher visibility and some measure of authority on the project. Those with an interest in management may become the chief interpreter and de facto high priest of product requirements.

When the engineering group writes the requirements document, it may take several exchanges between the engineering and marketing groups to make sure engineering accurately captured the requirements of the product envisioned by the marketing group. Sometimes this becomes an extended game of "guess–guess" where engineers guess at the requirements only to find they guessed wrong. In such a case one may ask why the marketing group doesn't write the requirements themselves. There are many answers to this:

- The actual details of the product may not have been determined. Interacting with the engineering team actually helps solidify the product concept.

- Various members of the marketing team may be genuinely unable to write a useful requirements document. The ability to precisely specify the operation of something with no inconsistencies or hidden assumptions is somewhat rare. In addition, the "do-the-deal" personality that allows a person to excel at his or her marketing job sometimes has associated personality traits that overlook discontinuities, contradictions, and mismatches.

- By the nature of their job, the members of the marketing team often suffer more interruptions and are on the road more than most engineers.

Aggressive engineers may demand that the marketing group write the
requirements and may try to force it to do so. This is generally a bad idea.
Important people in the marketing group, trying to protect themselves or
their staff, will likely characterize such engineers as hard to work with. This
will likely happen at meetings where the engineers are not present, and at
this point his or her career has been damaged. It is far better for engineering
managers—or one of the engineering staff who did well in English composition—to write the requirements document themselves. Not only does writing
the document allow you some amount of control of the requirements; everybody also likes it when you do work for them. Write the requirements document for the marketing group, and you will win the popularity contest.

At some point, all the involved parties agree that the requirements document is completed and correctly captures the essence of the product. This
should be consummated with a ceremonial sign-off of the final version of the
document. A sign-off by the senior representatives of each involved corporate division signifies their agreement with the content of the requirements
document. Without such a step the document may not be read by those not
directly involved in its preparation and the engineers have no guarantee of
a fixed target for their design work. There must also be a defined process for
amending the requirements document to accommodate changing needs. In
a just world, a change to the requirements should cause a reconsideration
of the schedule that was based on the original requirements. In the real-life
world of engineering, however, a very interesting phenomenon occurs. Often
the engineers are held to the original schedule even if significant amounts of
already completed design or implementation must be redone as a result of the
requirements change. I cannot offer an explanation for the logic behind this,
nor can I suggest a perspective where this might be fair to the engineers.

8

Plan the Effort

Introduction

Nothing of any complexity can be done quickly and efficiently without a plan. However, no matter how well we plan it seems that something always goes wrong. The most important thing about a plan is to understand that it is an imperfect device made by imperfect humans. Some flexibility is needed, and it is always good to have contingency plans for the most likely deviations. It is also appropriate to note that the plan does not need to be completed before starting the design. Likewise, the requirements need not be completed before starting the plan or the design. There can be a great deal of overlap in the project stages I've outlined in this book. Indeed, this is a book about career management, not about project management. This book describes high-level project concepts and discusses associated social interactions and how they may affect your career. These project concepts are presented as a tool to allow discussion and inspection of the engineering existence. To become expert in planning or managing a project you should take a course or read a book that explores the details of the structure and various stages of managing a technical project. From the perspective of this book, it is assumed that all the thinking and planning and all the strategizing and designing are done at relevant and timely points of the project.

The Project Plan

All this strategizing and planning is most beneficial when communicated among the project team members. Writing a project plan is a traditional and effective way of doing this. Although the specific content of the project plan varies from corporate culture to corporate culture, it generally covers a broad range of tasks and responsibilities spanning and defining the entire project. These may range from assigning personnel to arranging joint ventures with other corporations and from estimating production volume to handling the future obsolesce of critical components. More specifically, a project plan might include some or all of the following:

- A name, an overview, and a statement of the goals of the project
- Anticipated staffing requirements and identification of key personnel
- A definition of needed project skills, especially unique skills
- Selection and specification of a development environment including the number of development platforms and how they will be distributed among the engineers
- A specification of configuration control for hardware, software, and mechanical development
- A specification of error reporting and tracking for hardware, software, and mechanical problems
- Standard corporate processes and procedures to be used, waived, or superseded
- Certifications and approvals that need to be obtained from customers, vendors, or the government
- Security or conditional access issues that must be cleared with providers or government agencies
- Specialized test equipment that must be created or obtained
- Itemization of known intellectual property and royalty issues to be addressed
- Joint ventures and other contractual issues
- Performance, usability, quality, and environmental standards
- Risk identification and circumvention
- Links to various reference documents, the requirements document; the project schedule; and project hardware, software, and mechanical design documents
- Project marketing, sales, and technical budgets; cost estimates; and production cost goals
- Sample and preproductions plans including the number of development and debug platforms to be made, the number of preproduction units to be made, and the facilities and equipment to be used.
- Specification of mass production, including facilities, volume, testing needs, and a plan for transitioning from small-volume preproduction to full production
- Product warehouse, distribution, and delivery strategy
- Specification of the requirements change process

- The customer service strategy
- The warranty repair strategy
- Definition of staffing needs and a process for tracking and correcting product defects discovered during warranty repair or as a result of customer complaints
- Definition of a process and staffing needs for responding to component obsolescence problems during the production life of the product

Special Notes on Project Plans

It may require a great deal of work to produce a comprehensive project plan. For this reason, smaller companies may choose not to invest the needed resources. This might be the right choice for them. Usually, the projects undertaken by smaller companies are, shall we say, "smaller." In a smaller project fewer people are involved and the group is more tightly coupled. It is easier to identify critical responsibilities, there are fewer corporate divisions, and coordination among individuals, groups, and fiefdoms is easier. In addition, many smaller companies live or die on the hard work and talent of a few key individuals. In a successful small company, these experts might not have the plan written down, but they have it in their heads. They have thought through the various stages and needs of the project and know what has to happen when and who has to do what. These key individuals also have the ear and confidence of the corporate executives and therefore command the needed cooperation from those who must deliver segments of the project. In a large corporation, however, the divisional competition and political environment generally make it a good idea to write down the details of the plan and sign it in agreement. This eliminates (or at least reduces) disagreement about what group was responsible for delivering a particular milestone. It also forces organizations and agenda to at least temporarily focus on the same goal.

Project plans sometimes contain a great deal of confidential and proprietary corporate information. In these situations it may be a good idea to produce multiple "sanitized" versions of the complete project plan. These special versions are customized so they are suitable for distribution among different audiences to ensure that the project duties and responsibilities of all individuals and groups are communicated and understood. It is never a good idea to simply keep the project plan secret. People and organizations function best when information is shared and understood.

A Project Plan Is Not a Design Document

A college student or novice engineer may be quite surprised at how often major project decisions are determined by business and marketing considerations instead of engineering investigation and design. Major decisions such as the choice of the processor or operating system are often made based on business partnerships, co-marketing agreements, and legacy corporate loyalties. They are even made based on a perceived need to follow the current crowd of products rather than an engineering evaluation of the suitability of the latest technical fad. The project plan captures the stated business benefits of such selections and decisions. In this way the project plan is distinguished from design documents that specify the technical architecture, engineering trade-offs, and component criteria that were considered in the creation of the product.

The Project Schedule

A critical part of the overall project plan is the project schedule, and all good project schedules have a variety of long- and short-term milestones. Short-term milestones are assigned to components that constitute the long-term milestones. They are needed to allow a rapid determination when things go off course and, therefore, expedite corrective action or perhaps adjustment of customer expectations. Experience and judgment are involved in choosing the right amount of detail in the project schedule. If there are too many details, maintaining the schedule becomes a purpose unto itself instead of a tool to effectively manage the project—if there are too few, things can go dreadfully wrong before detection. A good project schedule will track issues critical to delivering the project and contain a relevant selection of the following dates and milestones:

- What engineers will be added to the project on what date?
- When must the requirements document be finalized and signed off?
- What are the completion dates for the initial design of the hardware and software?
- What are the completion dates for implementation of hardware and software subsystems?
- What are the completion dates for unit testing the individual hardware and software subsystems?
- When is the first hardware available that can be used by the software developers? Experience indicates that it is good to have the specific

requirements of that hardware written down and agreed to by both the hardware and software groups.

- When will test software be available to validate the new hardware platform's suitability for use by the software group?
- When will test software be available to validate that the hardware platform is complete and correct?
- When will hardware and software validation of the hardware be complete?
- What are the dates of successive integration milestones for hardware and software subsystems? There should be several as increasing complexity is layered onto the system. Big bang system integration is a bad idea because finding underlying bugs is much harder in the more complete and complex system. An incremental approach allows layering and discovering underlying bugs at an easier, less complicated time in the process when only a minimal number of potentially unfriendly subsystems are interacting.
- What are the dates of any needed demos?
- When is software implementation complete?
- What are the availability dates for special test or diagnostic equipment?
- What are the availability dates for subsystem test plans?
- What is the availability date for the system test plan?
- What is the date when the comprehensive system test starts?
- What is the date when the comprehensive system test completes?
- What is the product completion date?
- What are the availability dates for manufacturing test fixtures, test plans, and supporting test software?
- What are the dates for needed vendor or government certifications, approvals, and validations?
- What is the date of pilot and preproduction manufacturing runs?
- What is the date when final product mass production starts?

Special Note on Delivery Dates

It is worth distinguishing different kinds of scheduled delivery dates. A *hard date* is one determined by the physical world such as a needed planetary alignment. A *set date* is one determined by humans such as a holiday buying season. There are also *subjective dates* where someone has, for example, made an informed determination that the product must ship by a certain date to hit a marketing window or perhaps to receive funding from investors. Do not confuse the types of dates when making the product development schedule. Missing a hard date means that all the effort that went into the project was wasted. In most cases, there is no reason to do the project in the first place if is unlikely completion can be achieved by the hard date. In this situation, scheduling must be conservative, and it is best to use proven technology and proven developers. There is a little more flexibility with set dates. If a set date is missed there may still be future opportunities, or perhaps it just would have been "better" to hit the set date. Subjective dates are still more flexible and therefore allow a reasonable amount of intelligent risk taking.

Special Note on Project Schedules

Have you had to wait at home for a delivery only to have it show up late or not at all? Have you bought a house, planned settlement, and hired movers only to see the builder miss the promised delivery date? These are instances of events that happen every day where the responsible parties have years of experience with processes and designs that haven't changed in decades. If they can't get it right what hope is there of hitting a schedule where new technology has to be created to overcome the insidious conspiracies of nature?

Yet hope springs eternal. Engineering managers and corporate executives continually search for ways to motivate their staffs to stay on schedule. Note that this is slightly different from saying that they search for ways to produce an initial schedule that accurately predicts the ultimate project timeline. In fact, many projects on which I worked had little interest in creating a schedule that accurately represented realistic deadlines and contained contingency and recovery plans for likely delays. There are many possible reasons for this:

- Such schedules are often perceived as unnecessarily pessimistic by executives. Career advancement is most often awarded to those who create an image of being an aggressive can-do person. Even if the schedule maker is repeatedly proven wrong and overly optimistic, executives often appreciate the sentiment of trying hard rather than being a whiny defeatist offering a gloomy schedule.

- The project may not be approved if the "real" development time and effort is known. However, once the project is well under way and the company has already spent money it would be much more inclined to continue funding it.

- Some (most?) managers believe the engineers will not work as hard if they are not given a tight deadline.

- An aggressive schedule is received from another group (often marketing or business development), and the engineering project manager is effectively told to do whatever is necessary to hit that schedule.

- The schedule maker is blissfully unaware of his or her inability to make an accurate schedule and is continually surprised at how long things take. Often this is not due to ineptitude but rather lack of familiarity with new technology or lack of calibration of the staff. Staff calibration is really an under-recognized component of producing an accurate schedule. The number of years of experience a developer has may not be a good indicator of the speed with which an engineer will complete an assigned task. Some are faster than others, and some make more mistakes than others. Until you work with someone for a while it is difficult to gauge how long it will take that person to perform a task.

Although some projects did not emphasize making a realistic schedule, every project on which I worked expended significant management effort to track progress, recover schedule slips, and motivate the project staff to work harder. One of the great frustrations of engineering life is a continuing succession of projects where you suffer the self-fulfilling failure of overly aggressive schedules. In some cases where the aggressive schedule was intended to be a motivational tool, the pressure-driven work ethic backfired as frantic engineers took increasingly desperate design shortcuts or implementation risks to meet their deadlines. Like betting heavily to win back gambling losses, these risks and shortcuts often snowball, causing the project to fall further behind schedule. Sadly, the project management feedback loop results in more pressure as the project falls further behind. A project that lags behind schedule can spiral into chaos as more pressure, more status meetings, and a variety of increasingly desperate recovery plans distract the engineers from the real work of investigating and solving problems. Specifically, with years of experience on a multitude of diverse projects I have never seen an aggressively scheduled project recover once it has fallen significantly behind. I have, however, seen career-damaging disasters caused by zealous project recovery plans run amok.

Following are some general rules that may help to increase the likelihood of developing a product on a predictable schedule:

- Use proven developers.
- Use proven technology.
- Copy a previously successful design for a similar product.
- Hold groups and subsystems individually accountable. For example, it is inappropriate to compress the software schedule to compensate for hardware's being late. It is also inappropriate to compress the integration test schedule because everything else was late.

- Ensure test equipment is obtained before it is required.
- Order more development licenses, emulators, and debug boards than you think you will need. Development "stuff" is cheap. Engineers and time are expensive.
- Protect the developers from the distractions of supporting previous projects.
- Avoid excessive full-team meetings.
- Build test and debug tools; build lots of them.
- Finally, do your engineers and the project the biggest favor: Finish the requirements document, sign it off, never touch it again, and make sure everyone on the project knows what it says.

Special Note on Manufacturing Testing

I often find that intelligent and experienced people are surprised by the level of difficulty in getting a smoothly flowing high-volume production line. It is so much easier to build one of something that works than it is to build large volumes of something that works. There are, of course, many issues with qualifying components, vendors, and designs. What I'm referring to is the difficulty in designing and implementing a production line test system able to rapidly and reliably test large volumes of a product. The test process of high-volume manufacturing is not concerned with proving the design is good. The goal of a manufacturing test is to ensure that a functional product was assembled correctly from good components. A variety of sophisticated optical scanning techniques can be used to determine if all the components were installed correctly, but the hardware also has to be designed to allow connection of somewhat invasive test fixtures.

Ultimately, many of the platform's functional tests depend on activation of software resident in the product's code. Some of this software may be used only for manufacturing tests, and some may be part of the normal operation of the product. In either case, this software can be written only with detailed knowledge of the product platform. Writing the test software just cannot be successfully accomplished without the cooperation and support of the product development team. Moreover, some of the more complex manufacturing test software functions can be efficiently written only by the expert engineers intimately familiar with the design and implementation of the product. This burden on the product development team is often not anticipated or scheduled, and this unexpected overhead can be responsible for large schedule slips. An independent manufacturing test software development team can create the user interface and can write the bug tracking, report generating, and other modules involved in validating the manufactured product. They can even call test-function application programming interfaces (APIs) created by the product development team. However, without training that effectively incorporates them into the development team, the manufacturing team cannot write test software that actually runs on the platform. Everyone will be happier if this necessary distraction of the platform development team is anticipated and accommodated.

Special Note on Nurturing the Manufacturing Process

It is reasonable to expect a complicated manufacturing process to experience growing pains. One should expect a poor yield from the first, or perhaps even the first few, production runs. It is best to ramp up the production volume in stages to allow kinks to be worked out with smaller quantities first. Development and debug units generally provide a good early opportunity to build small volumes of the product. Limited volumes of preproduction units for a beta test provide another opportunity. Unless your product is a rare exception, some problems will afflict early attempts at mass production. Solving these problems often necessitates technical experts visiting the manufacturing facility to see the problems first hand. Complex products may require that multiple experts make numerous trips to the production plant. This can become very expensive if the manufacturing is being done 8,000 miles away. The envisioned savings of the cheap but distant labor quickly evaporate with the overhead of even moderate production problems. Unfortunately, the expense of debugging and resolving these kinds of problems usually comes out of the engineering development or support budget causing a smudge on the career of the responsible engineering manager. This extra cost may not show up on the manufacturing budget, allowing the executives who made the decision to build the product on the other side of the world to reap honors and rewards for their money-saving choice of manufacturers.

The hidden support cost of distant manufacturing is one issue, but there are also problems with schedule delays and perhaps staff morale problems that result from dozens of trips around the world. Unless other arrangements were made, all of these problems automatically become those of the product's engineering development manager. Budgetary, scheduling, and other responsibilities for the support of the manufacturing effort should be assigned in the project plan to avoid contention and misunderstanding when manufacturing problems predictably occur. Having a few problems in the transition to volume production is pretty common and should not be a surprise. Intelligent anticipation of problems allows one to avoid surprise, which is fundamental to creating and maintaining an accurate schedule.

Project Plan Sign-Off

The project plan is the ideal place to capture all the previously discussed topics and perhaps many more. However, the best and most comprehensive project plan does no good if is not distributed to all the involved parties and if there is not a commitment from those parties to the correctness and viability of the plan. My experience is that most projects are late simply because the initial schedule is unrealistic or overly aggressive. Final sign-off of the project plan is more than a formality. It is an agreement of all involved that the plan is viable, achievable, and not excessively aggressive or optimistic. Once such an agreement is reached there must also be accountability. Everyone makes mistakes and accidents happen, but there must be a justification if signers of the project plan are unable to deliver on their commitment. Understanding why a commitment was missed is not from a need to blame someone but rather to create a learning process so each successive project benefits from previous experience. Over time there must be fewer and fewer surprises and therefore progressively more accurate and predictable schedules. Over time managers will also establish a history, and some will repeatedly miss deliveries due to unforeseen events.

When individuals, despite coaching and the opportunity to learn, continue to be surprised by projects, they are demonstrating a lack of anticipation and vision. Eventually this lack of vision must disqualify them from future project planning. In contrast, some managers will demonstrate an almost uncanny aptitude to predict problems and even the approximate amount of time needed to solve them. Such individuals are to be revered, as their vision enables accurate schedules and therefore efficient assignment of corporate resources. In a well-run company this efficiency translates directly into bottom-line profitability. A realistic schedule may not be as politically popular as a more optimistic one, but there is bankable money in accurate project schedules. Capable managers working with signed-off, high-quality project plans can ultimately create accurate project schedules if the politics, posturing, and mindlessly excessive optimism are not allowed to destroy the process.

9

Begin the Project

We have an idea, we know it's a good one that will make money, and we have a solid plan in place. What comes next?

Assemble the Project Team

It's time to start building the product, and to do that we must assemble the development team. Ultimately, this team will be responsible for completing a broad variety of tasks, from selecting components to writing the user manual and handling warranty repair. However, not all members of the team will be needed throughout the project. A quality project schedule will enumerate the needed team members during various points of the project. Certainly designers are needed early in the project as are experts in configuration management and project management. As the project matures, more implementers will be needed. Later, the number of implementers will begin to decrease, and the number of system integrators and testers will increase.

Once the product is shipped, the project team will consist of a support staff handling problem reports, manufacturing issues, and warranty repair.

So exactly how does one go about assembling needed project personnel? In a large company there may be previous projects that are being completed or perhaps a few people who weren't very busy. In any size company you may have to hire new employees or bring on contractors. A common problem is that some of the folks will continue to be distracted by old projects that might not have finished on time or perhaps by spontaneous catastrophes that need immediate attention. Such delays and distractions are ubiquitous and afflict nearly every project. A well-done project schedule anticipates them rather than blissfully assuming they do not exist. Getting the project off to a delayed start is no way to achieve success. The real challenge is not in pulling together the project team on a timetable compatible with some schedule but in doing sufficient homework to create an initial project schedule that correctly anticipates the delays and distractions always associated with assembling a focused project team.

Sometimes engineers may be "lent" to a project from another group. This is actually a fairly common practice because few managers want to see a reduction in the number of people they supervise. A large head count reflects positively on the manager, and at some point, as the count grows, a promotion is in order. Permanently reassigning staff results in a decreased head count and may indicate a negative career path to some. However, an engineer on loan to a project may experience a mixed allegiance, especially when the lending manager continues to provide direction to the engineer that distracts from work on the project. This can be a serious problem for the project because the engineer must favor the person doing his or her review when resolving conflicting direction.

Be aware that a snag sometimes occurs when acquiring project engineers from another group. Managers rarely give up their top performers willingly. Be cautious of engineers made available from groups whose managers share no responsibility for the success of your project. The right thing for an individual manager sometimes supersedes the right thing for an extended group. A coworker of mine had an engineer working for him who was not very adept at his job. At an executive staff meeting another manager expressed a dire need for additional staffing for his project. My coworker volunteered that he had a guy who just finished a task and, in the interest of being a corporate team player, he was sympathetic to helping the group. It would hurt, he said, but he would make this individual available to the manager. A few weeks later I asked the "dire need" manager how things were going. The manager observed that the new helper was "a little slow." The moral of this story is that managers fiercely retain their best people. Be wary if you are freely offered an unknown person from another group. The mantra of keeping a project on schedule is to use engineers with whom you are personally familiar. Unfortunately, this is not always possible.

The Blame Game

Other sociopolitical problems occur when new people must be hired for a project. Here I'm not talking about the obvious risk that no matter how well the interview went the new person might not work out. I'm talking about more insidious career risks that happen either accidentally or as a result of clever manipulation by a politically astute rival. A personal incident that illustrates this problem occurred when the manager of a group running behind schedule interviewed some candidates and offered a position to one of them. The manager then asked me to phone interview the candidate only days before he was to start. This put me in an extremely difficult situation. If I declined to interview the person I would be viewed as unsupportive and not a team player. If I interviewed the person and did not give a positive recommendation, the manager could correctly say that I had rejected the person and was therefore responsible for delaying the hiring of a needed resource. Delaying a resource then would have associated my name with the continuing delays in that group. Even if I approved the person I would not be involved in supervising or directing him or her. However, my name would still be associated with any delays because a resource I approved was unable to do the work in a timely fashion. I would share in the blame even if the inability to do timely work was caused by poor management of the group. How does one solve such a conundrum? I decided to gracefully avoid the problem by being too busy. Unless some high-priority crisis is involved, it is difficult to criticize you when the corporate demands of your job make you too busy to immediately jump into an additional task.

Was I just being paranoid? Not with the manager involved. Over time you learn to calibrate your coworkers and to determine which of them conducts their life by Machiavellian principles. Some people place self-advancement above all else. Image is everything, and an image tarnished by mistakes or delays is less likely to get the next promotion. Certain people just cannot abide this, and somebody else must always take the blame and the fall. There are enough such people that some amount of paranoia in the workplace is advisable. My engineering career advice is to be cautious and to look for ulterior motives and agendas. In some cases they exist.

Personality, Personality, Personality

Is there some philosophy that can help pick good developers? Is there some discernible quality that good development team members have in common? There is no magic bullet, but if you named the three most important attributes of a valuable product development team member, the answer would be personality, personality, and, yes, personality. Sure, having really smart people on the team helps, but only if they have a personality that allows them to make contributory use of their intellectual gifts. A really smart person whose number-one goal is self-advancement has no place on my team. A really smart person who is careless by nature can do more accidental damage than he or she can ever compensate for with talent.

What personality attributes should be sought when selecting members of the development team? Perhaps the team should be populated with the oft-requested and highly prized team players? Perhaps not. An occasional problem with team players is that they wait for direction instead of attacking a problem on their own initiative. Some managers find this desirable. For them, *team player* is a euphemism for people who do what they are told. Populating the development staff with some number of team players is unavoidable, but the critical developer positions benefit greatly from aggressive team-leader-type personalities. They attack problems, actively seek out coworkers to gain information, and demand that vendors deliver on promises. Sometimes this ruffles feathers, but team-leader-type behavior can yield rapid progress on projects.

Progress can be even more dramatic if multiple members of the group exhibit this behavior in a cooperative fashion. Be sensitive to the fact that some managers have difficulties with this type of project environment because it involves activities and decisions not under their direct control. Such difficulties can result from good intentions (the managers worry about the quality of the decisions), or bad (they are insecure or perhaps demand to be in the project spotlight). Some amount of infighting and intellectual competition may occur when several members of the development team have the team-leader assertive take-charge attitude. This need not be a bad thing if the development manager has the temperament to mentor and direct these aggressive personalities toward benefiting the project. Success at this generally requires strong managers who empower their people to make decisions and who have sufficient technical competency to maintain the respect of these aggressive staff members. It is critical that bickering and competition not become destructive, and the managers must monitor their team and genuinely lead it to ensure that interaction and competition among the players maintains a constructive quality.

The positive side of populating a project with aggressive team-leader personalities is that it can improve project quality and shorten the delivery schedule as few other techniques can. Are there negatives? Sometimes aggressive and talented people may get the reputation of not playing well with others. Such personalities can be detrimental to a project, but not playing well with others must be carefully distinguished from not playing well with idiots. I've known people who don't play well with anybody, those who don't play well with idiocy or insanity, and those who don't play well with subordinates. Managing those who don't play well with anybody is a painful chore. Managing those who don't play well with subordinates may be superficially easy because such people may flatter and praise those above them. Be watchful of treachery, however, as such flattery and praise are usually intended for self-advancement and are not sincere. If your roles were reversed, you would be mistreated along with the other subordinates. Finally, managing those who don't play well with idiots can be a delight, so long as you are not one of the idiots. If you do not possess technical talent it can be difficult to know good advice and suggestions from bad and therefore difficult to distinguish idiocy from genius. It is not automatically assumed that people who tell you good news and support your instincts and decisions technically know what they are doing. It merely means they have good interpersonal skills and know that their future success and advancement depends on keeping you happy. It may be hurtful to your ego, but often those who argue with you and tell you you're wrong are the ones you should listen to the most. Understand that those who argue with their manager are either too socially inept to understand the consequences or feel strongly enough about the issue to risk future promotions. Waving your magic supervisor wand does not fix every problem. If intelligent members of your staff are arguing with you, they are putting their future at risk, and you owe them the courtesy of trying to understand their issue. If you are technically competent, you will have no problem managing a development team full of aggressive team-leader personalities. If you are one of the idiots and don't like taking advice from your subordinates, you would be much better off building a development staff from team players.

Self-confidence is perhaps the most important facet of the team-leader-type personality. It takes self-confidence to believe you can track down and fix complicated problems and even more confidence to believe you can fix somebody else's problems. Confident people tend to have debugging styles that differ somewhat from those of their less secure teammates. Tentative developers may approach a bug with the mindset that they must fix it by making only some minor patch or tweak to the existing implementation. While often true, this viewpoint tends to intellectually limit the options considered by the engineer and may slow or even prevent rapid resolution of more difficult problems. Confident engineers simply have more degrees of freedom in their thinking.

Describing other personality traits consistent with rapid development of high-quality technological products includes words like *passionate, dedicated, committed,* and *hardworking.* As a group, these words convey an essential attitude that the engineer believes in what he or she is doing and is determined to see the project to successful completion. One aspect of such dedication could be described as stubbornness, or perhaps more generously, tenacity. Another aspect amounts to something of a discontent with the status quo, an unhappiness with mediocrity, or simply a striving for greatness. It has been said that all progress begins with malcontents. In a sense this is very true because the content person sees little reason to change or improve something. A great engineer is one who combines passion with innovation. These engineers are the type of people who get annoyed about having to wait at a traffic light for a left-turn arrow when no cars are in the left turn lane. They are unhappy with inefficient and incomplete products and have the passion to push the design to that next level and make a product great. These passionate malcontents are incredibly valuable and perhaps even indispensable to speedy development of innovative, high-quality products, but they can be a challenge for bureaucratic managers. Passionate malcontents want to make a product of which they and their company can be proud. However, they may view formalities such as frequent status meetings, diluting features to maintain a schedule, and repeated prioritization and reprioritization of bugs as impediments to rapidly shipping a good product. Not every company and organization can abide such viewpoints. Organizations, like individuals, have distinct personalities and cultures, and passionate malcontents may not mix well with stodgy or autocratic environments or bureaucracies where contented managers view preserving the status quo as paramount.

One final personality attribute is critically important to being successful as a design engineer: pessimism. An engineer responsible for the design and correct implementation of a complex product must be instinctively pessimistic, skeptical, and distrustful. It is the engineer's job to be so. Any engineer who takes action based on unproven assumptions, self-delusion, or wishful thinking is not worthy of the title of engineer and would be better off in a different career. While a marketing or political campaign may be successful with mindless optimism and malleable facts, engineers must overcome problems in the blemished and uncertain real world where truth cannot be negotiated. Helen Keller said: "No pessimist ever discovered the secret of the stars, or sailed to an uncharted land, or opened a new doorway for the human spirit." This is an inspirational statement, but the reality is that some pessimism is also needed for all these things. Pessimists know that opening a new doorway means that one must be prepared for what is found on the other side. Pessimists must be in charge of building a ship that can travel to the stars, otherwise a starship built on optimism, hearsay, and invalid beliefs would never leave the launch pad.

Pessimism, however, is the ultimate double-edged sword of an engineering career. Managers and corporate executives enjoy a Pollyanna's blithe assurances that everything is going well much more than the engineer's report of impending disaster. Pollyanna quickly gains the reputation as a positive influence and a can-do type of person whereas the engineer is viewed as Chicken Little and is relegated to the laboratory where he or she is viewed as a difficult to work with negative person. The reality is that the death, doom, and disaster engineer is the corporation's best friend. An engineer's job is to anticipate the things that can go wrong and avoid them. It can be difficult for managers and executives to know all the problems anticipated and avoided by a good engineer because those problems simply never happen. Pollyanna, however, continually encounters unanticipated problems and through great fanfare overcomes them. In many corporate cultures, Pollyanna's repeated demonstration of problem-solving ability reinforces the view that this type of person is a highly valuable and exemplary employee. Busy and technically unsophisticated executives may not realize that a more pessimistic person would, through skill and vision, have far fewer troubles to overcome. In this way, the engineer's career becomes schizophrenic. By using their talent and experience to foresee and predict looming problems engineers risk becoming unwelcome corporate pariahs. Sometimes it is tempting, very tempting, to feign being blindly optimistic and then to respond quickly and decisively to problems you knew were coming and could have avoided entirely.

Leadership, Trust, and Talent

Sometimes project managers are not technically sophisticated but sincerely want to do what is best for the project. In this case, managers must be able to identify knowledgeable people to trust. This can be very difficult for managers who don't personally understand the technological issues. The reason for this was well described by Arthur C. Clarke, when he said, "Any sufficiently advanced technology is indistinguishable from magic."

To one not versed in the needed science it can easily seem to be magic. To the untrained, those who wield technology can seem to be magicians who speak in gibberish. How can well-intentioned managers identify whom to trust when they can't even ask good questions? Unfortunately, the most reliable method of identifying skilled engineers takes the longest. Over time, good engineers will establish a history of correctness. Note that this is subtly different from a history of success. It is entirely possible that a highly talented engineer could work on a succession of project disasters when the astute advice and recommendations of engineers are overruled or ignored by management or other engineers. The important observation is the ultimate correctness of their observations and suggestions. At times a good memory (or perhaps written notes) is needed as others attempt to rewrite history to take credit for good ideas or simply to avoid having been wrong. Of course, the entire concept of tracking the ultimate correctness of various engineers is impossible if managers never seek or listen to advice directly from the engineers.

Correspondingly, less gifted engineers can be identified over time by observing those who repeatedly encounter unexpected delays and problems. Anticipating technological problems requires no crystal ball, only some skill and the vision to apply previous experience to a new project. There will always be excuses to explain why certain problems were not foreseen. Such glitches may include components' being unavailable when promised, holidays of overseas cultures that shut down some element of production, unexpectedly difficult tasks, and staffing problems. Sometimes problems are genuine surprises that no one could have anticipated, but often engineers are inexperienced, not looking ahead as they should, or have observed that the "Pollyanna" has received promotions and more favorable treatment than the "Chicken Little." The latter case is a fact of life in many corporate cultures, which it makes it even more difficult for managers to determine whom to trust.

Over a shorter time frame it may be possible to identify good engineers by the company they keep: Who are their friends? With whom do they eat lunch? As in high school, where socially ambitious individuals sat at the lunch table with the cheerleaders, there will always be some hangers-on trying to improve their status by associating with the talented engineers. Some observation, however, can easily determine the key folks and who is gravitating to whom.

For nontechnical managers, the worst way to identify good engineers may be to judge them by how much they tell you what you want to, or perhaps expect to hear. Socially aware people recognize that their superiors control their future financial comfort, and this provides a strong motivation to adapt information and to tailor reports to put events and the project status in the best possible light. To overcome this embellishment and to determine the truth it may be necessary to ask questions and challenge assertions of good news and rapid progress. This is especially critical for marginally technical managers who may not understand quite as much as they believe.

Project managers need to lead the project and to set expectations. Lacking an understanding of the project's underlying technology puts managers at a significant disadvantage but does not prevent them from holding people accountable for their schedules and the quality and completeness of their work. Successfully designing and implementing a complex technological project is not about eliminating features, sacrificing test time, and executing progressively more desperate schedule recovery strategies; it is about finding problems, beating them into submission, and then finding more problems. Anybody can build a bad product, but it takes a good technical team with good project management supported by good corporate stewardship to make a great product.

Communications

Building a capable development team is a challenge, but making and keeping it productive means establishing and maintaining effective communications among the team members, consultants, vendors, supervisors, marketing, sales, testing, manufacturing, and others involved in the project. For project managers, enabling communications is an important job. They should circulate the project plan, requirements document, and schedule among all parties. If some portion of the documents needs to be confidential, they should create "sanitized" versions of the documents that can be freely distributed to ensure no information critical to a group or company has been omitted from their version. It is poor project management if someone on the team hasn't seen documents relevant to his or her activities. Project managers should clearly articulate the method of revising and redistributing the various documents so there is never an excuse for designing, implementing, or testing the wrong thing.

Define terms. For example, when something is said to be done, it means that it is finished; you never have to touch it again, and you can deliver it to the customer with pride. Something is not finished when the initial development has been completed. It must be integrated into the overall system, tested, and most likely have bugs fixed. Careless use of terms like *done* creates a misconception of the true status of the project. It is also important to ensure that the relevance of schedule events is understood. For example, when a milestone indicates a completed subsystem, it really means that it must be delivered to the testing group weeks earlier. Simple misunderstandings of terms and events can interfere with the successful flow and progress of a project.

Often a company's hardware and software groups exist under separate management hierarchies. In environments with poor communications this can create multiple problems with the architecture, design, and ultimate viability of a product. The reason is simply conflicting agendas. Hardware success tends to be evaluated against criteria such as low cost, high reliability, and high manufacturing yield. At a fundamental level, all of these measures are related to minimizing the cost of reproducing or repairing the product. Minimizing the product cost either directly or indirectly biases the hardware design toward less memory, a slower processor, and only hardware that is absolutely necessary. Rapid development of high-quality software, however, benefits from plentiful memory, a high-power processor, and extra hardware to assist the software in complex tasks.

Overcoming these fundamentally conflicting needs demands tight communication between the hardware and software teams. Any hardware architecture that is not reviewed and accepted by knowledgeable software developers virtually guarantees schedule delays or worse. The expertise needed to properly approve the hardware architecture is specifically that of knowledgeable software engineers, not managers or hardware designers. This is because the software team members are ultimately responsible for making the product work, and they must participate in the ownership of the design. I've seen numerous underpowered products delayed for months while the software engineers struggled to squeeze needed functionality or performance from the hardware platform. This is okay if the company has made the conscious commitment to gain a market advantage by using economical hardware, but it must be communicated that the necessary tuning and optimizing of the software is not free and is paid for in both schedule delays and software engineering expenses.

Another area that benefits from close communication between the software and hardware teams is the use of a new or unusual hardware design or new components such as a custom application-specific integrated circuit (ASIC). There are many good reasons for such hardware migration, but, again, this activity needs to be coordinated with the software team to ensure that necessary development and debugging environments exist for the new hardware.

All the good reasons for new hardware mean relatively little if the software folks can't efficiently write and debug the software. Even with adequate development and debugging environments the hardware and software groups must closely coordinate testing and initial activation of new components. Special test boards and platforms may be needed and special (throwaway) software may be required to evaluate and debug new hardware. Availability of suitable test platforms and this additional workload of the software and hardware teams must be considered in the schedule. Of course, a delay in the delivery of working test software is owned by the software group, and a delay in availability of working hardware is owned by the hardware group. In an uncertain and mutually developmental environment, neither group should be allowed to claim that the other is slowing overall progress.

The hardware and software organizations also need to communicate with nonengineering groups in the company. Sometimes features or components desired by marketing or the executive staff have serious adverse consequences, but there is a natural tendency to interpret a request or an exploration by senior staff as a directive. In some cases important features are omitted because implementation will delay completion. There are also instances where the intention of saving money has unexpected negative consequences. For example, I'm familiar with a product that used a small-footprint real-time operating system (OS). This OS had a royalty cost under $1 per box. Senior executives viewed this as a wasted dollar and asked that a popular "free" operating system be used. Unfortunately, the new OS had a much larger footprint that necessitated adding several dollars' worth of memory to the product. Although various members of the technical staff were aware of this, it was never communicated to the executives.

Engineers and engineering management can be faulted for being poor communicators, but people must also be willing to listen. Some organizations concentrate on making a product with the lowest possible cost. Others focus intently on maintaining a short development schedule. It is not about rapidly producing a cheap product; it's about making a successful product. There are times when products ship with defects and when competitive features are omitted. Sometimes this occurs due to poor communications on the part of the engineers, but there are also times when such problems occur over the protests of the project engineers. Engineers must communicate certain facts to management, and management must be willing to listen when the facts are presented. The corporate organizational tree, method of computing bonuses, and the hierarchy of accountability must be structured to accommodate this reality. The alternative is that competing priorities and agendas allow the continued delivery of feature-deficient and low-quality products to market.

Establishing and maintaining effective communication within the development team can be a challenge for the project manager, but this takes on a far higher degree of difficulty when some of the team is remotely located or scattered around the world. Most crippling is the lack of many of the unofficial communication channels such as lunchtime conversations and walking over to someone and asking how he or she did something. After observing multiple companies struggle with numerous attempts at offshore development of embedded systems, I've arrived at the conclusion that such development inherently has limitations. First, bringing up new hardware and new software at the same time is not readily compatible with shipping frequent hardware revisions through customs around the world. Second, things work much more smoothly if the tasks assigned to the disparate groups are clearly separated with a clearly defined interface. The management dream of the sun never setting on development stumbles in the reality of daily handoffs of dynamic engineering activities between the groups. My observations here are specific to embedded systems where there is significant interaction between the hardware and software teams. Success in this environment requires experienced program managers skilled at organizing tasks, motivating employees, and establishing clear lines of communication.

E-mail is a wonderful method of communication but also has a dark side. Occasionally e-mail conversations go on for days without resolution. It is so easy to copy additional people and to jump in with comments that e-mail conversations sometimes diverge instead of reaching a consensus. Project managers are responsible for clearly setting project expectations. It is reasonable that e-mail be limited to a few exchanges; then pick up the telephone or call a face-to-face meeting. Also, it is much faster for senior people to whip off an e-mail to subordinates than for subordinates to carefully craft an intelligent-sounding response to their superiors. Each e-mail is cheap, but the cumulative cost of extended technical discussions or a barrage of butt-covering and posturing e-mails can be high. Some moderation in e-mail traffic can dramatically improve project communications and productivity.

Ultimately, the amount and format of project communication depend not only on the personality and management style of the project manager but also on the corporate culture. Some managers like to have all communications go through them, and some are okay with direct engineer-to-engineer communications. Some managers choose to maintain control and to restrict dissenting opinions by limiting access to documentation and information. Some corporate cultures tolerate this. It has often been said that knowledge is power. The corollary is that ignorance engenders cooperation. It is worth noting that some cultures and managers are more interested in control and self-advancement than in honest exchange of information, designs, problems, and solutions. For some projects this is just an unfortunate fact of life that impairs communication, diminishes morale, and slows development.

Properly Equip the Team

Engineers are expensive. From a variety of perspectives, anything other than the engineers' salary is a relatively insignificant part of the development cost.

Acquiring the best development environment, the best debug environment, and the best test equipment positions the team for success and maximizes the productivity of the expensive engineers. When ordering hardware prototypes for development, configure them with maximum memory. Automate processes where possible. Not only is it expensive for engineers to do mundane and repetitive tasks, but, being human, they also make mistakes. Enable home development. Engineers come to work because they get paid to do so, but many of them play with technology every waking hour. Giving them a development system at home can result in significant acceleration of the project effort and maybe some unique new features.

For consumer and many commercial products, properly equipping the team also means having them personally using early versions of the products. So many bad products would never be shipped if those responsible for them actually had to use them. Like a scratchy tag in the back of a T-shirt, you have to marvel at some things on the market and shake your head. Didn't they ever try to use it? For the executive staff, using a product means pulling it out of the box and installing it yourself. I've often seen crippling product installation problems hidden when an engineer or technician goes to the executive's house and installs it. Rule 1: Where possible, allow the developers to use the product as one of the customers would and feel their pain. Rule 2: Take the time to fix the usability or quality problems before shipping the product.

Design First

This is arguably the most important stage of the project. A good product starts with a good design, and poor execution at this stage can doom the project to a prolonged and painful recovery. A bad or omitted design can result in higher costs, lower performance, or complete and utter failure of the entire project. The necessity of a good design is blatantly obvious, but what makes a good design?

A design isn't some doodles on a sheet of paper, and it's not some high-level theoretical work. Because the design must actually be built it should not have options and choices. It specifies what is to be built and how to build it. A good design does not have significant "to be determined" areas, and it doesn't have big holes in the middle and hazy clouds that roughly approximate some kind of magical happening. The first step in creating the design is to put down the soldering iron and push away the keyboard. The next step is to think, and the step after that is to think more. It is at this point where blank paper paralysis is most often encountered. Overcoming blank paper paralysis can be very difficult for some engineers. Others can sketch out important aspects of the system architecture relatively quickly. They can rapidly identify which subsystems or operations must have priority and where bottlenecks are likely to occur. Because of this highly visible difference in engineers' aptitude for creating a system from a blank sheet of paper, some gain a reputation as expert designers. Management may view them as too valuable to remain on the project after the initial design is completed. Unfortunately, too many so-called experts create a design that is difficult to implement or that overlooks certain requirements. Sometimes fitting new or neglected requirements into the existing framework is quite difficult. Removing the original designers breaks a critical continuity chain with the remainder of the project. Keeping the original designers on the project to help implement it enforces accountability for the design and promotes distribution of the details and intentions of the design. It also speeds changes and modifications that may be needed as the project evolves.

Though it's true that a good design doesn't have significant unspecified areas, in many cases necessary information is simply not available at the start of the project. In such circumstances, maintaining a reasonable schedule necessitates overlapping implementation activities with a continuing design effort. Sometimes the unknown areas are critical to the overall project, and considerable technological investigation must be conducted to fill in the blanks and mitigate risk. What this really means is that the engineers don't understand some things and have to figure them out as the product gets implemented. I specifically distinguish lacking information to complete a new design from the previously discussed Pollyanna-like failure to foresee problems regularly encountered during the course of building a new product. A Pollyanna is unable or unwilling to see future difficulties in the normal and customary flow of a development effort. Lacking specific information to complete a design is a completely different issue. The missing information could be understanding limitations of a component, learning a new technology, needing some lab experiments, input from customer focus groups, or any combination of these and other investigations. In a complex product there may be some number of iterations of designing, implementing, learning, and redesigning. It is an unfortunate fact of engineering life that understanding an unknown can require an indeterminate amount of time. When project managers hear that the engineers don't understand, are investigating an issue, or are redesigning some feature, the managers are usually terrified of possible schedule impact. There is good reason for them to be concerned. If the unknown areas are critical to the overall architecture of the project, delays in resolving the issues can result in substantial and unpredictable project delays. Drawing on experience, engineers can estimate about how long it might take to figure things out. But these estimates are merely educated guesses and can be wrong—very wrong. Worse, as the solution is unknown, there is little that can be done to contain the delay other than work harder, work smarter, or find additional expertise.

As common sense would indicate, delays are most common on new or unknown technologies. There is little excuse for unforeseen delays when copying an existing design or building a new version of an old product. However, the same product functionality does not necessarily imply the same internal design. Creating a cost-reduced second generation of a product is not a simple task if you change, for example, the processor, operating system, bus structure, device drivers, data compression, or encryption schemes. Something as simple as changing the processor speed can yield unexpected and difficult-to-fix timing problems. A product may work the same and look the same from the outside, but a mountain of new technical problems must be solved if the internal workings are altered significantly. Unfortunately, it sometimes requires a skilled engineer to determine when the internal workings have been altered significantly.

Design delays can also be encountered as engineers discover that layer after layer of additional research is needed. This is most common when building a new product on the "bleeding edge" of technology or with a technology unfamiliar to the engineers on the team. In this case the engineers may be no better than laypeople at estimating the complexity and difficulty of designing and implementing something with the new technology. Capable but unfamiliar engineers may think they are close to a solution but repeatedly discover additional investigation is necessary. Management frustration can increase with each unexpected setback, and the frustration can build to aggravation as continuing delays erode the perceived marketing window. The aggravation is understandable; however, some problems are really hard, and even the best engineers are only human. Berating engineers or beating on them like a schoolyard bully doesn't help and only causes an antagonistic relationship and poor morale. Unhappy engineers may not be outspoken enough to challenge those who pay their salaries, but they are certainly thinking, "If management is dissatisfied with our progress, they should solve these problems themselves." Unhappy engineers could also be thinking, "I need a new job," and losing a good engineer is always damaging to the company.

Throwing a design together and hoping for the best is not generally a path to success. It is better to make a design that simply must work. Unfortunately, when a design that must work is actually implemented it usually doesn't work right away. This is normal and expected for things created by humans. The key distinction is that it is expected to work and that the engineers therefore have a vision of how it is supposed to work. This provides a logical debugging path to follow to diagnose and correct problems. It is here that an important attribute of good engineers comes to the forefront. Good engineers insist on understanding why something that should work doesn't. They don't start making random changes to make the problem go away. They logically walk through the design, examining intermediate states and computations and looking for where the actual implementation first deviates from the design expectation. When they find the problem, they fix it.

A philosopher might observe that a major difference between scientists and engineers is that scientists are excited by unexpected results whereas engineers abhor them. Scientists are excited because they may have stumbled onto some new mystery or nuance in their investigation of the universe.

Engineers detest unexpected results because they are trying to build a product on a tight schedule and don't really have time to deal with the whims of the insidious universe that continually conspires against them. Unexpected results indicate that some unknown factor is interfering with expected operation. Unknown factors must be tracked to their source to be sure the design and implementation will work well in all circumstances. One of the worst things that can happen is when an apparently irrelevant change makes a problem go away. People who pretend to be engineers are happy the problem is gone. Genuine engineers dislike this because something is happening that they don't understand, and this is usually a bad thing. Experienced engineers also dislike this because they know the problem is not really gone—merely hiding. It will come back later when more features have been added and the system has grown in complexity, making the bug all the harder to find. All engineers earning their paycheck will track down why an apparently unrelated change made a problem go away to understand why it happened.

Finally, some design decisions are really business decisions and not the result of technological evaluation. Choosing a CPU for the product is one example where the dominant factor may be the business relationship with the CPU vendor rather than an engineering assessment. From a business perspective the key question is how does this choice of processor improve the market opportunities for the product? Perhaps the CPU manufacturer is providing free software or offers support that could help produce the product and get it to market quickly. Perhaps the manufacturer is well known and offers to share in its large marketing budget. In the corporate environment the challenge may lie more in figuring out how to use a processor chosen for business reasons rather than selecting the best one for the job. Unfortunately, cost-sensitive products rarely have processors with an abundance of power. Often the question is not whether the CPU can be used but how well the product can be made to work with the one already chosen and how much software tuning will be needed to get a reasonable amount of performance.

Determining a CPU's limitations by analysis prior to implementing the design can be very difficult. One of the first things needed is to identify the usage scenario that yields maximum processor demand. It should be fairly easy for someone familiar with the product to identify the various operations that must occur. What is much harder is to understand how much CPU needs to be devoted to each of these operations and which ones may make demands on the processor simultaneously. Each of the simultaneous operations (e.g., reading from a DVD and playing a movie while the user is pushing a button on the front panel) involves a certain amount of labor that must be done in a very short period of time. However, the exact amount of that labor depends greatly on the underlying contributions of the hardware and the quality of the design and implementation of the software. Specifically, the design and implementation of the software can always be made bad enough to prevent any processor—even the most powerful—from working well. Before the load on the processor can be estimated, something must therefore be known of the software design and expected implementation. The software operations that must be performed simultaneously add additional complexity to this investigation. In reality, operations on the same processor do not happen at the same time. To the human, they appear to be simultaneous, but the software is really performing the tasks in a sequential fashion by loading a descriptive context for each one and working on it for a short while before loading the context of the next. As the number of tasks that must be done in a short period of time increases, the amount of effort devoted to these context switches grows nonlinearly. At some point the CPU cannot cycle through the needed tasks fast enough, and the game becomes creating a design that gracefully degrades instead of crashing.

Early testing of cost-sensitive designs with marginal processors should be expected to expose situations where the processor saturates and the product misbehaves. This is not a case of poor engineering but of real-world trade-offs associated with making the product more economical. Time must be allowed for optimization of the software to eliminate the most egregious misbehaviors. Such system tuning can be time consuming and therefore makes development more costly. The hope is that the cheaper hardware cost of the marginally powerful CPU gives the product an advantage in the marketplace and enables a volume of sales that more than offset the higher development costs. At some point a senior person in the corporation decides that the imperfect product actually works well enough or that enough money has been spent on development. In either case, it is time to ship the product.

Determining if a CPU will work is only one example of a large number of business and technological decisions that must be made during the design of a new product. Real engineers do not let preconceived notions dissuade them from the truth. Indeed, as 19th-century scientist Thomas Huxley said, "The great tragedy of science is the slaying of a beautiful hypothesis by an ugly fact."

It is common sense that a faster disk drive is needed for proper operation of products that heavily use one. However, as part of designing a product I once looked into methods of selecting a hard drive fast enough for our application. Testing a variety of hard drives on the market indicated that the fastest was maybe 20 times (2,000%) faster than the slowest. That was interesting and exciting information. However, additional investigation determined that usage of different file systems, operating systems, device drivers, and implementations could affect performance by more than 400 times (40,000%). It turned out that the actual hard drive was not very important compared with several other critical design choices. This was so counterintuitive that it was viewed as blasphemy by the project manager. Despite the facts, a fast hard drive and a marginal file system were selected for the project.

Many, many product design decisions must be made. Some are technical, and some are strictly for marketing purposes. Some decisions are the result of business pressures, and some are architectural. How big does the product have to be, how much can it weigh, how much memory is needed, what kind of security is necessary, how much power will it use, how hot will it get, and how long does it have to last? No wonder blank paper paralysis is such a common problem. Most importantly, the design stage of the product is the time to be creative in addressing the technical, marketing, and business issues involved in getting a high-quality product to market quickly.

10

Manage the Development

Introduction

Management is a force multiplier. Through the hands of others, managers can accomplish the work of ten, twenty, or hundreds of people. If managers are righteous and true and know what they are doing, the project has an excellent chance of experiencing prosperity and success. If managers are clueless or evil, darkness can descend upon their project and minions. It is important here to understand that *managers* here mean the people really in charge. A figurehead manager being manipulated and ordered by others is mostly unimportant to the success of the project. It is the person ultimately in charge who determines success or failure.

Education and experience are important, but a great deal of project management is hard work and common sense. Hundreds of books on project management and numerous university and industry studies refine and present techniques and schemes for successful project management. These references are valuable, but in my experience projects generally don't fail because people don't know what to do. They fail because the correct things just didn't get done. There may be any number of reasons that it seems like a good idea to circumvent established processes, to ignore a specific risk, to delay addressing a serious bug, or to cause disruption and chaos in a headlong drive to hit a specific milestone. Much of this damage is caused by good intentions gone astray, such as wanting to please a superior by reaching an important milestone. Other damage may come from simple human failings such as vanity and ambition. The core of project management is very simple. People and events need to be organized in a coordinated effort toward the goal of successfully completing the project in a timely fashion. To accomplish this, project managers need a realistic plan—a specific order of tasks and sequence of events, based in reality, that ends in successful delivery of the project. The absence of such a plan amounts to industrial roulette where the success of the project is left to chance. If the manager's engineering plan is to "be lucky," he or she should be playing the lottery, not managing the project.

Management is the culmination of the previous project stages. At this point, the requirements should be frozen, there should be an accepted and signed project plan, and a talented development staff should be in place.

I've seen organizations try to save money by hiring cheap implementers and having an expensive and talented manager ride herd on them. That doesn't work. You can't manage brilliance into a turnip. Bad implementers can do more damage more quickly than managers can intervene and correct. There should also be a design in place or one that is coming together before significant implementation begins. Nearly everyone talks the talk; few walk the walk. Lots of people embrace creation of a design before implementation starts, but overzealous or misdirected managers sometimes demand that "real work" begin before enough of a design is in place for the developers to intelligently begin implementation.

As with the medical profession, an interesting mantra of project management might be, "First do no harm." Managers who can help solve technical problems are an asset. When project managers merely think they can help solve technical problems, they can be a distraction or worse. Managers who avoid making decisions or who continually change priorities or shift resources may soon be managing a disaster. Project success requires the steady guiding hand of a manager who stays focused on executing the plan and has the vision and foresight to avoid common and predictable pitfalls. Capable and experienced project managers know that every new requirement, every demo, and every interim or partial product release delays final completion of the product. They understand that the natural order of the world is to first make the product work and then to make it work better, faster, and more reliably.

While it is true that every release and every demo delays final delivery of the product, there is no avoiding the reality that multiple releases and demos will occur during the life of the project. The so-called demo tax and release tax must be paid, but the fee can be minimized with advanced planning. Releases and demos should be scheduled in such a way that required features and performance are compatible with the normal development path. All hardware and software used for demos and interim releases should be configuration managed and version tracked so that quirks and observations can be logged against a particular version and later reproduced.

Requirements Are Your Friend

At the fundamental level a requirements document exists to tell the development team what to build. It also exists to act as a device to restrain change. Often, either because a product was not initially well conceived or because of turbulence in the marketplace, requirements must change. A comprehensive project plan should describe a change control board that reviews requested requirement changes. Although some changes are relatively minor, it must be understood that every change comes with a price. Engineering must be allowed to reasonably adjust the project schedule every time the requirements change. Depending on the new requirement, there can be a significant project impact as changes ripple through the design and various subsystems of the product architecture. Test plans and user manuals may also need to be updated, and customer service agents may need to be retrained. In summary, it's a bad idea to be changing the project once you start designing it.

Configuration Management

Early in the project some form of configuration management must be instituted. There needs to be a way for everyone on the project to be able to find the latest copy of critical documents. Don't even think about launching software development without source control and the ability to track changes and label builds so that versions and results can be reproduced as development progresses. Some forms of software configuration management are quite slow and greatly extend build times over what can be done on individual PCs or workstations. In many circumstances it's okay for individual developers to keep private copies of the code to speed iterative build and test scenarios. In such cases, there must be regular backups and a well-defined code reintegration procedure. However, you must ultimately depend on the intelligence and good work habits of the developers. Developers who lose more than one day's work due to a PC or hard drive crash should be placed in stocks in front of the office and publicly humiliated.

Configuration management and associated processes should be designed to help the developers achieve robust and repeatable results without getting in their way. Prior to starting a new project the marketplace should be surveyed to see if new and better tools have come available. Rebuilding and linking all source modules might have worked when a software project consisted of 50 files, but it might be much better to consolidate subsystems into separately managed libraries when thousands of files go into a project. Never be afraid to creatively review processes and approaches for possible improvements.

Motivating the Team

Shortening the Schedule

It is pretty much universal consensus that the project development team needs to be motivated. As you may recall from earlier portions of this book, my experience is that most good engineers went into engineering because it was fun for them. In addition, I've found that most engineers, especially the intellectually elite, love the technical part of their work. They try hard and work long hours, especially when they are chasing a difficult (i.e., fun) problem. Many also "play" at home on the same or related technologies for no pay. So you have intelligent people working hard and sometimes even on their own time. How can you motivate them even more? One universally accepted method is to create a project plan with an absurdly short development schedule. This can be done in many ways. One is to simply state with authority that there is a limited market window that must be hit. I was never really sure what this meant, but it is generally stated with such authority that nobody challenges the premise. Amazingly, this authoritative premise happily coexists with the well-known marketing principle that it is rarely the first to market that ultimately succeeds (think IBM PC or Microsoft Excel). Nevertheless, someone has somehow determined that a product must get to market by a particular date, or sales of that product will be compromised. The short schedule is then used to encourage the engineers to work harder. Unfortunately, short schedules sometimes have the inadvertent consequence of also motivating the engineers to take silly or high-risk shortcuts. This is especially dangerous when project managers are not technically adept and can't readily determine when such dangerous shortcuts are being taken.

From a philosophical prospective it's interesting to think about three different companies' presenting their development teams with the exact same "market window" for competing products. Two of those companies rush products to market. The third company finds some sanity and slows the headlong dash to get to market quickly. Sadly, the most likely scenario is that the rushed products suffer from the shortcuts used to get to market quickly whereas the "late" product benefits from superior fit, finish, and quality. Which would you buy?

Working Smarter

There can be more to motivating engineers than trying to get them to work harder. You can challenge them to work smarter. Dinosaurs were very strong and powerful, but the sneaky little humans took over the planet. Maybe humans had a little help from a comet, but our smarts gave us the ability to make spears, to invent the wheel, and to learn to control fire. As humans seized control of the earth from powerful competitors, working smarter and producing better products are the secrets to seizing market dominance from corporate competitors. When you think of all the difficult to use, unreliable, or just plain bad products on the market; achieving market dominance by "producing better products" must be a well-kept secret indeed.

Challenging engineers to work smarter reinforces the very reason many of them chose their profession. Rising to this challenge requires some amount of fearlessness on the part of engineers. Not all feel comfortable going against the status quo or in trying to enhance a proven design. Some with thick skin have no problem posing difficult questions or alternative approaches. Most worry (with good reason) that such behavior might irritate their superiors and jeopardize future promotions. For the greatest benefit, creativity must be nourished and encouraged. Engineers exist to invent and create. Expedient solutions and silly shortcuts are not natural or beneficial to the profession or the product. John W. Gardner, 1964 recipient of the Presidential Medal of Freedom, had the following observation about encouraging creativity:

> When Alexander the Great visited Diogenes and asked whether he could do anything for the famed teacher, Diogenes replied: "Only stand out of my light." Perhaps some day we shall know how to heighten creativity. Until then, one of the best things we can do for creative men and women is to stand out of their light.

Tangible Recognition

Trinkets and food and awards, oh my! Many companies have some means of rewarding engineers for a job well done with a nominal amount of cash or corporate stock. Several also encourage employees to be thinking about patents by rewarding them with cash for patentable ideas. The amount of money is generally not enough to change the lives of the engineers, but the goodwill and encouragement that come from the recognition can very well engender a more positive relationship with the company. Likewise, giving out corporate coffee cups and shirts or other useful items is a great way for the company to demonstrate it is gracious and supportive. The importance of a positive relationship between the corporation and the employee cannot be overemphasized. Employees can easily pretend to work hard while accomplishing virtually nothing. There are times when hard work is needed and expected of the team. At all times, however, engineers and other staff members should be encouraged to continually look for ways to make their job easier and more efficient.

It's a nice gesture for the company to provide food when the team works late into the evening on a project. Likewise, donuts for morning meetings or cookies for afternoon meetings are economical ways for the company to create goodwill. Project T-shirts are a good way to establish camaraderie and a sense of family on a project. T-shirts done with taste and minimal graphics can be worn for years and also prove to be an inexpensive corporate advertisement and recruitment aid.

Unlike a manual laborer or assembly-line worker, it is difficult to know when engineers are working hard and fully devoting themselves to solving a problem. Determining the level of commitment is doubly hard for managers with minimal understanding of the technology involved. Engineers who feel abused or unappreciated may find it extraordinarily easy to get by with minimal effort. Motivating engineers, like most people, involves treating them well, making them feel wanted, and demonstrating good leadership. So much about motivating the engineering staff is simply about not demotivating or alienating them. Great effort comes naturally for engineers when doing something they love in a challenging and supportive environment.

Positive Reinforcement

Cheerleading is a necessary part of corporate executives' jobs. Everyone would worry if the boss were to say the project or company is in trouble. Seasoned executives know this and understand their position of influence. They very rarely present a negative image and simply must say positive things to effectively lead the company and to avoid discouraging the staff. However, the executives must walk a fine line when attempting to motivate the staff working on a difficult project. If there is too little cheerleading, a sense of fatalism or a negative image is conveyed. If there is too much, the developers who know the true state of the project may think the executive is out of touch or posturing. The corporation depends on hard work and goodwill of capable developers to make current and future projects a success. Loss of their respect damages the relationship between the corporation and critical employees and seriously damages their commitment to the company. Engineers simply don't work as hard if they think the corporate leadership is clueless. The best way to keep them engaged and committed is to acknowledge difficulties and to set realistic goals. Sometimes it is very difficult for certain executive personalities to restrain their bravado and to communicate honestly about problems and difficulties with a beleaguered project. Doing any less, however, risks the commitment of those most critical to the future success of the company.

Vendors and Subcontractors

The Houdini Test

As a showman, Harry Houdini would call on members of the audience to place him in handcuffs. This greatest escape artist of all time would then amaze the spectators by quickly removing the most secure and complex devices. Early in his career he encountered a scoundrel who had poured sand in the lock mechanism, making the handcuffs impossible to unlock. Although Houdini had to be cut out of the handcuffs it was a beneficial experience because he learned that not everyone was trustworthy. Thereafter he demanded that handcuffs be locked and unlocked in his presence before they were secured on his person. He never again suffered the embarrassment of having to be cut out of someone's handcuffs. Houdini's success came not only from talent but also from astute rejection of tricksters and charlatans. Demanding that a vendor show its product working before it is integrated with your product is called the Houdini test. Many vendors exaggerate the capabilities of their products to some extent. Sometimes the exaggeration gets out of hand and escalates to outright lies. When a vendor's component is integrated with the overall project it can be much more difficult and time consuming to track down quirks or missing functionality in the vendor's implementation. Worse, an unscrupulous vendor can be less than supportive in finding bugs. Delays can become extensive when determining whether the vendor's product is bad or simply not being used correctly. It is best to always have a clear specification of the capabilities expected of a project component delivered by a vendor and a way to independently validate conformance to that specification.

Working with the Vendor

At times, a vendor greatly overcommits to delivering some functionality. As milestones are missed, the level of overcommitment becomes progressively more obvious. The easiest and most often used method of managing the delinquent vendor is to simply demand that it delivers on schedule. Further delays gradually escalate the issue through layers of management, culminating in corporate executives' yelling at the vendor's executives. Realistic or not, the most frequent result is that the vendor's executives agree to institute a recovery plan, to add more resources, and to do whatever else is necessary to promptly release the needed functionality. Unfortunately, the vendor's executives often lack detailed knowledge of the actual problems causing delays, and their primary goal is to manage the irate customer. Their commitment to fix the problem may simply be to pacify the customer for a few more days until they can deliver something to quiet their client. In this situation the customer did not do himself any favor by backing the vendor into a corner and demanding delivery. An honorable vendor was already trying to deliver on its commitment. A dishonorable one could care less about the customer's concerns or demands and will continue to delay delivery or perhaps will deliver junk as a stalling tactic.

From a project perspective, it matters little whether the vendor's overcommitment was innocent or malicious. Either way, the project is delayed. Simply demanding that the vendor deliver is not a path to success because it is based on blissful ignorance of the real cause of the delay, and blissful ignorance is never a dependable engineering strategy. It may seem that ordering the vendor to deliver is managing the issue and forcefully taking charge of the problem, but this is an erroneous belief. Although you have clearly placed blame on some other organization, a successful project is not about finding someone to blame for failure but about achieving success. It is better to work with the vendor to understand its problems and to help out. In the worst case, this provides a mutually trusting working relationship and a more accurate view of the true schedule.

Core Expertise

Technology companies have core expertise they consider essential to their survival. The company maintains a high level of proficiency and investment in these technologies to ensure that its capabilities are second to none. As such, development work in the area of core expertise is generally done by employees of the company, not vendors or subcontractors. Personnel changes, advancement of technology, and evolving customer preferences may gradually erode the company's core competency and its relevance. Over time, it becomes necessary to investigate and learn new and unfamiliar technologies. The way a company goes about updating its technical capability is a reflection of the corporate culture. Some companies make it their business to be at the forefront and to help create new technology and new applications. Others fear the new technology will take too long for their busy engineers to learn, so they seek a development partner with the needed expertise. Occasionally the new technology is "bleeding edge," and the partners must work through the kinks and difficulties together. In a partner scenario, the company to some extent has lost control of its own destiny. The quality of the engineers and the priority applied to solving problems is under the control of another company. Without a careful legal framework the eventual working technology, paid for in the blood and money of both partners, may become the property of only one. What could have become a new core competency instead became royalty payments to a now necessary partner.

The decision to contract out development in a new technology is a difficult one involving many factors, including the apparent complexity of the new work. However, a great deal of so-called new technology is actually an evolutionary, albeit innovative, usage of something existing and understood. Because of the novelty of the usage, the barriers to entry sometimes appear formidable but fall quickly as experience is gained. The seemingly daunting barriers may cause a company to contract out the development when additional investigation would have determined it better to add it to the company's core competency. Although such outsourcing may not be especially good for the company, it occurs in large volumes at numerous organizations and provides a continuing source of funding for new small companies. Some of these small specialists will use this revenue stream to become large mainstream companies. The process repeats itself year after year as the little companies become larger and themselves contract out advanced development as they become more concerned with stability and predictability than with innovation.

There are also times when a company contracts out development work in a new technology because it has become frustrated with the progress of its own engineers. A few missteps or a couple of project disasters, and corporate executives may become concerned about their engineers' ability to learn the new skills. In reality, the problem is more likely to be learning new things on an aggressive fixed schedule. Busy corporate executive sometimes lose track of the engineers' need to explore. Learning a new topic in school amounted to learning enough about it to correctly answer a few questions. Mastering a new technology well enough to build a quality product on an aggressive schedule requires a certain comfort and fluency with the technology. Smoothness and grace in applying a new technology of course requires research but also some amount of unguided experimentation and modeling of various usage scenarios. When viewed in this light, it is not surprising that a crash course in a new technology often leads to a crash of standard development projects. Project disasters with new technology are not only understandable but also predictable if the engineers only studied enough to answer a few multiple-choice questions in their headlong rush to ship an important new product. The core competency of a company comes from a number of engineers' being intimate with that technology and having a deep understanding of its nuances. Adding a new technology to this list requires some patience and restraint. Such a step is better as an exploratory or investigational research project where engineers are allowed time to acquire a thorough understanding of the capabilities and shortcomings of this new tool without having to justify daily why they are behind schedule.

Design Reviews

Just about everyone will agree that it is correct practice to have a design in place before implementation begins. Having a design is important, but having a good design is even better; one of the best ways to assure that the design is good is to have capable engineers review it. The design review is an established method of improving the quality and shortening products' time to market. Although this is proven, known, and understood, it sometimes seems like a good idea to skip this step to save time and to avoid distracting critical designers and implementers. In many ways the apparent time savings achieved by keeping the engineers focused on continued design and implementation is erroneous. Several hours can be saved today by skipping the review, but recovering from design oversights or mistakes can cost weeks or months later in the project. Early in my career I worked with a brilliant man who taught me it was best to not have to apologize for a design. The design review is one of the best ways to assure that little apology is needed and that development is headed in the right direction. If there is no time for some portion of the staff to invest a few hours in a design review, the project is already behind schedule and will likely get worse. This bears repeating. If there is no time for a design review, the project schedule is already in jeopardy.

Though the exact format of a design review may vary, successful ones have a few things in common:

- The review should be done by your best people. Like saving the best corn for replanting, sharing the knowledge and experience of your top folks can help other designers get better with every project.

- Although the reviews should be done by your best people, it should not always be the same engineers doing reviews. This avoids always burdening the same people and also allows others the experience of participating in the review process.

- The design should be distributed prior to the design review meeting to allow time to examine and understand it. Distributing the design at the meeting pretty much reduces the design review process to correcting punctuation and grammar.

- Make sure those reviewing the design understand the importance of this job. It is not to be taken lightly, and there must be adequate preparation so a real contribution can be made to the quality of the design.

- Make sure those whose design is being reviewed understand that some discussion and criticism are to be expected. They must take this as constructive criticism and not view revisions to their design as damaging to their careers. Along these lines, as much as managers may want to be present, attending the design review should be done with discretion. Reviewers may hesitate to criticize engineers in front of their managers, and the engineers being criticized may react defensively instead of being receptive to the comments.

- Actually fix the design as needed. Reviewing a design is good, but time must be taken to further investigate potential problems and to correct genuine flaws. Not bothering to actually fix discovered problems could result in a catastrophe. Not bothering to actually fix discovered problems can have serious adverse consequences.

It is also beneficial to have implementation reviews. This not only eliminates obvious errors but also may improve the overall quality of the implementation. The general quality may get better because implementers are likely to be more careful if they know their work will be examined by others. Many bad or deficient implementations happen because implementers feel rushed and take shortcuts. Taking the time to perform implementation reviews sends the clear message that heavy schedule pressure does not justify bad work. With software, reviewing the implementation achieves a significant portion of the observed advantages of pair programming at a very small fraction of the overhead. I'm not a big fan of pair programming because the overhead is enormous and the alleged benefits can also be achieved by conventional good management and an occasional implementation review. Pair programming, like socialism, can help the weak folks but greatly hinders the better ones. Two idiots working together can't mentor each other. Two capable programmers don't need to mentor each other. In my mind, pair programming only makes sense if the manager can't distinguish the incompetents from the gifted programmers.

Pyrite Engineering

Skipping design reviews and doing "quick and dirty" implementations are examples of a general class of project problems that may be collectively referred to as pyrite engineering. Pyrite is also known as fool's gold because it can be mistaken for real gold by uninformed people. Pyrite engineering consists of cutting corners and taking shortcuts with the intention of shortening the project schedule. Like fool's gold, this of a false value. The project always pays for shortcuts and stupid engineering tricks. Payment can be made today by investing in good initial work or later when the project manager has run up an intellectual deficit and borrowed against the project's future by demanding quick and dirty implementations. The real value—the real gold—is to do the job right the first time. Do it right, and you will never have to do it again. Do it right, and you will encounter fewer project delays later down the road.

Be in Charge

There is the person who is really running the project, and then there may be a figurehead manager who has no real authority but does a great deal of the work and takes any necessary blame. If you are the real project manager you are the face of the project and the one who does presentations to customers and senior management. If you are the real project manager no meetings happen in your absence where decisions are made about your project. A former coworker was once the project manager of a major project but was too naive to realize that he was only a figurehead manager. Many decisions were made about his project in his absence. Sometimes the decision makers even forgot to inform him of the decisions. His orders were often countermanded, and he was frequently instructed to do things he considered detrimental to his project. One telltale sign that he was just a lackey was that he prepared the project status report, but his report was edited and presented to senior management by someone who worked in another division. The real revelation occurred when a vice president forbade him to make a presentation at a customer site with the reasoning that he was "too senior" and had grown out of that role. Those reporting to him needed to do the presentation. Interestingly, one of those reporting to him who got to present at the customer site worked in the vice president's division and was also the person already reporting to senior management on the project.

If you are nominally in charge, insist on actually being in charge. The worst that can happen is you get fired—but you probably didn't want to work under abusive or demeaning circumstances anyway. It's okay to delegate presentations and related matters if such activities are not well suited to your nature, but this should be your decision. As the project manager you should have the right to do the presentations, and the content of the presentations should be under your control. If you are forbidden or overruled, at least you know your real position in the organization and you can plan your job future accordingly.

Make sure everyone understands that it is not appropriate for project meetings to happen or for project decisions to be made without you. It is not always possible to be at every meeting, and there are some meetings you don't want to attend but always need to know about them. At best, holding meetings related to your project without your knowledge is rude, and can be an indication of much worse organizational or political problems. Also, there are simply no project decisions without your participation. There can be a tentative decision, but it is not final until you concur with it. Discuss the problem with your superiors if meetings and decisions continue to occur without you. Insisting on the right to run your project is definitely aggressive and can make enemies. However, someone else who is taking credit for your effort or working behind your back has already proven he or she is not your friend.

Occasionally, the excuse is made that you are too busy to do the presentations (or attend meetings or make decisions). This is asserted most forcefully when your project is behind schedule. The reasoning is superficially sound. The project is behind schedule, and you need to focus on that—not do presentations. There is some judgment necessary, but if it is your project to run you must be allowed to run it and must be allowed to represent the project to senior people. There are many methods to help you when you get behind schedule on a busy project other than giving the project recognition and visibility to someone else. Perhaps a presentation could be delayed, or perhaps someone could be made available to help you gather status or to solve problems as needed. The important point is that the person running the project should not be invisible or ignored.

From the corporate viewpoint, a stand-in for the designated project manager should be viewed with suspicion. The executives may question why they are getting secondhand information from someone other than the designated project manager. Is this personnel switch an attempt by some other supervisor to get "face time" for one of his or her people? Is there some reason the real project manager is being hidden? The corporate executives have every reason to be skeptical and suspicious if the designated project managers are absent from meetings and presentations and if decisions are being made without them.

Teflon Management

Even with a good plan, good staff, and good design, things can go wrong. In every project things go wrong. Sometimes things go a little bit wrong, and sometimes they go really wrong. When things go really wrong, corporate politics generally necessitate finding someone to blame. Unfortunately for naive engineers, there are plenty of brilliant politicians in the corporate workplace. These schemers are very adept at deflecting blame onto unsuspecting engineers who are more interested in solving cool problems than in who gets blamed for delays and catastrophes. Solving cool problems is why most people become engineers, but it is not career-safe for engineers to drift through the work environment unconscious of the Machiavellian ambitions of those around them. Deflecting blame for problems and delays becomes instinctive for some who aggressively seek career advancement. Unfortunately for socially unsophisticated engineers, the deflected blame lands on them, tarnishes their star, and dims their chances for future promotions.

Solving cool problems usually means difficult, challenging, or meticulous work or perhaps inventing new technology to do things never done before. All of this loosely translates to high risk. For an engineer, there is an unfortunate direct and significant correlation between a fun project and a high-risk project. The upwardly mobile bureaucrat knows it is usually not a good career move to be associated with a project that is likely to fail. Should a high-risk project succeed, it is much easier and less risky to simply take credit for its success. Perhaps it was a skilled hand of shrewd management or the brilliance of initially selecting the right people that allowed the team to triumph over adversity. There are many ways to insert oneself into the project limelight after it becomes a success. Engineers, however, are in the project at the ground floor and suffer the lost sleep and high stress of trying to build complicated things on a fixed schedule.

How can senior corporate management determine when people are ducking a risky project or trying to insert themselves into the limelight of a successful one? It is actually quite difficult for many reasons. Corporate executives are generally very smart but may lack specific knowledge of the technology and difficulties of a given project. They may not have the time or the inclination to delve into the details or learn much of the science. A nearly universal executive approach is to have a few trusted advisors who provide information on various topics. Being human, these advisors may make assumptions or depend on information from their trusted advisors. Being human, there may also be some partiality and shading of facts and events to benefit favored subordinates. To put this in engineering terms, the system runs open-loop. That is, each successive level of management relies on a few people from whom it receives information and advice. There is little cross-checking of the information. It is a rare corporate executive who walks around the lab and talks to the working engineers about their projects. Indeed, such interaction might only cause confusion if the executive lacked sufficient technical knowledge to understand the issues. As most companies are run by business majors, it is easy to see why technical problems and difficulties on projects get distilled and simplified as they move through each successive level of management. Nuances of complicated technical issues may become lost in smiles and handshakes by the time the people who really matter hear of a problem. In such a scenario it is relatively easy for Teflon managers to avoid responsibility for problems or slide into the celebration of a successful project.

Schedule Delays, Status Reporting, and Visibility

It is easy to report status accurately when your portion of a project is going well. However, if you start to fall progressively further behind and report this truthfully, you will become the focus of more stressful and ever-increasing attention from management. Falling behind schedule on a project is not especially good for your reputation and career advancement. You can develop a variety of negative reputations that range from simply being inept to unable to handle the pressure of a big project. With the career-damaging adverse consequences of missing a schedule being so great, the obvious question is, "Why not start a project with a leisurely schedule?" Good question. Unfortunately, the engineering group is often handed the required schedule and told to figure out a way to turn out the product in the allotted time. Whether the required schedule is determined by market demand, by the desire to ensure that the expensive engineers work hard, or simply by a promise made to a customer, the engineering group rarely has the opportunity to compose a leisurely schedule.

Why do projects fall behind schedule? The reasons are legion. Perhaps the engineers knew the schedule was overly aggressive but didn't want to argue about it. Perhaps they knew that the schedule was overly aggressive and did argue about it but were told the schedule just could not change. Perhaps the schedule was initially okay but bad luck with a risk item delayed the project. Perhaps there were unforeseen events, loss of critical personnel, distraction due to resolving problems with legacy products, or maybe just bad initial estimates of the needed work. So many things can go wrong that the real question isn't why so many projects fall behind an aggressive schedule but how one can rationally expect all the stars to align to keep a project on schedule when there is little or no margin for error.

Some project scheduling methodologies attempt to force an error margin, or buffer, into the schedule by incorporating two estimates for project tasks. One estimate is a conservative or safe estimate to which the developer is comfortable committing; the other is aggressive but possible. As you weave the various tasks and estimates together, buffers are inserted at appropriate points of the project with the intention that things can go a tad wrong but the project can stay on schedule. Variations of this technique are something of an attempt to legitimize the time-honored scheduling approach of adding a little padding to some of the estimates. Armed with complex scheduling algorithms and books and papers that provide written affirmation, the engineers and program managers can scientifically justify to corporate leaders the need for padding the schedule with buffers. Unfortunately, this strategy and the entire science of scheduling algorithms suffer from the same common vulnerabilities. All the strategies and algorithms are meaningless if the technology is so unfamiliar to the engineers that estimates are merely wild guesses or if corporate management simply mandates the delivery date. The reality is that sophisticated scheduling algorithms, multiple estimates, critical chains, critical paths, heuristics, propagators, buffers, dependencies, predecessor identification, and the like all fail if the engineers have no idea how long something is likely to take or if corporate management is unwilling to accept how long something is likely to take.

Things rarely go wrong on projects with familiar technology and proven developers' using known tools to accomplish a well-understood goal. Unfortunately, most real-world projects have a number of unknowns and uncertainties, and this equates to risk. Projects don't fall behind schedule because the wrong scheduling technology was used but because the schedule was too aggressive or something unexpectedly went wrong during the project. On a fundamental level, advanced scheduling science does little more than attempt to surmount oversights and excessive optimism on the part of the engineers, program managers, and corporate leadership. Skilled and experienced project managers inserting some padding where things usually go wrong can do a fine job estimating and tracking a project schedule with the simplest of tools. Unfortunately, they may not be able to justify to impatient supervisors why such padding is needed in their project. History and experience of such a need count for little when executives persist in demanding the shortest possible schedule. History and experience provide no proof that the shortest possible schedule is actually unrealistic. Expensive and complex scheduling technology may provide some justification and leverage to push the schedule out against the desires of reluctant management. Advanced scheduling tools do not shorten the product time to market by helping to solve technical problems; they only provide justification to management of a longer than preferred development cycle—one that can accommodate the normal and expected turbulence in the completion of project risk items.

The most sophisticated scheduling science in the world can't help if identified risky items fail or if poorly understood technology becomes finicky. A complex development project using risky or poorly understood technology provides all the makings of a difficult and explosive situation. Perhaps program management, corporate management, or the engineers either didn't realize or refused to accept how risky the technology was or how poorly it was understood. How exactly a project gets behind schedule may be irrelevant to the dimming of one's career prospects. At some point it becomes obvious to smart and aggressive people that there is no way to keep the project on schedule. For them the game becomes minimizing career damage. There is an old joke about the guy who confounds his friend by putting on a pair of running shoes when they are chased by a bear. The friend says, "Why would you do that? You can't outrun a bear." The answer, of course, is you don't have to outrun the bear; you need only to outrun the friend. In the corporate world this means that if a project is behind schedule one need only to outrun a coworker to avoid trouble. More importantly, you need not genuinely outrun a coworker; you need only present a reality-bending but persuasive status that indicates you are on schedule. Like teenagers playing "chicken" with their cars, multiple groups and individuals may indulge in "schedule chicken," where all deliver fabricated statuses that send schedule elements careening toward each other. As with the car game, the goal is to be the last to swerve away from a career-damaging collision. The one who swerves last can maintain the image of being on schedule.

Schedule chicken may help the careers of those who successfully hide their delays, but it generally has adverse consequences for the project. Sustaining a fallacious image of punctual delivery of schedule elements often involves making or perpetuating a number of costly and inefficient decisions. Maintaining the image of being on schedule may necessitate paying additional fees to expedite orders, overtime pay for various workers, payment for delivery of interim or partial implementations, and multiple purchases of successive revisions of components. A great deal of money can be saved by simply waiting for the final and tested service or component version to be completed. But this is just money, and sometimes the expense is justified if the money accelerates final delivery even a small amount. The real damage to the product occurs when adhering to the fanaticized schedule necessitates living with a bug, skipping a worthwhile feature, or incorporating unfinished subsystems into the final product.

Schedule chicken is a serious problem in aggressively scheduled projects. Calling the bluff of those who purport to be on schedule may require specialized knowledge of details of the intentions and usage of the delayed schedule elements. I once saw a group indicate as complete the milestone of delivering a development board to a software group. When delivered, there was no processor on the board and it was completely useless to the group, which caused them to fall behind in development of needed code. When the leader of the software group objected to the statement that the milestone was completed because the board had no processor, the head of the other group countered with, "This is no time to be changing the requirements." The software group leader had lost. Continued argument about the topic would necessitate getting into details none at the meeting cared about and might present the dreaded "difficult to work with" negative images to executives in attendance.

Managers who have any amount of meaningful interaction with the developers should know fairly quickly when tasks are falling behind schedule. They may choose, however, not to report this information in status meetings. They may have the sincere belief that they will be able to recover lost time in the near future, or they may be hoping for a miracle. They may also choose not to report the fact that they are behind schedule for the very practical purpose of avoiding pain and extra work. When managers report at every status meeting that they are behind schedule, they look bad and have to answer more questions than those who are on schedule. If they are falling further and further behind schedule every week, they look inept, have to answer lots of questions, and are asked to put together a recovery plan. The pain and extra work can be avoided by continuing to report good status until immediately before a public milestone that would have exposed the tardiness of the development. Immediately before being exposed, the managers announce one big schedule slip and simultaneously announce remedial action to correct the problem. Amazingly, this works pretty well. Of course, these managers get yelled at and have to go to a few meetings, but the total pain and additional work are far less than if delays had been reported for several weeks. In addition, the managers disarm many criticisms by proactively presenting a schedule recovery plan. Indeed, most such recovery plans fail, and usually such a project continues to fall further behind schedule; however, skillful and perhaps somewhat deceptive status reporting can keep managers and their projects out of senior executives' crosshairs.

What of accountability? How can blatant fabrication of status reports succeed in the corporate environment? Such deception is not as hard as one might think. People, including corporate leaders, are much more willing to believe things they want to hear or things that agree with preconceived notions than jarring information that forces them to see the world in a new and perhaps unfavorable way. In this sense, positive status reports of success are much better received and far less likely to be challenged than reports of delays and failures. Furthermore, bold use of schedule chicken and other, shall we say optimistic, status reporting techniques may be difficult for members of corporate management to detect if they lack a sufficient knowledge of technological and project details. At times, the only way such manipulation of project facts could be known to the executives is if someone directly challenged the validity of the status report in their presence. Who might mount this public challenge? The only folks likely to have access to conflicting facts would be subordinates actually working on the problems in the reporting manager's group. The hands-on working engineers, technicians, and testers are not likely to be at an executive status meeting, and even if they were they probably value their jobs too much to challenge their boss's statements in public. The very position of corporate leadership makes the job of leading the company more difficult. Most people will go out of their way to avoid antagonizing powerful executives and will do everything in their power to present good status that makes them happy. Sometimes the desire to present positive and encouraging information degenerates into illusion and misdirection. At such times the senior executives must rely heavily on their intuition and little else to determine a course of action.

Demos and status reports are ways for corporate executives and nonengineering organizations such as the marketing group to get visibility into the inner workings of a project. Participating in status meetings and demonstrations lets nonengineers feel as if they are involved in the process of developing the product. Positive status reports or a well-executed demo gives them confidence that the project is headed in the right direction and relieves their anxiety. In contrast, a bad demo and missed deliverables make them apprehensive and nervous. Highly stressed executives and marketing folks are no fun whatsoever. Few of them can help solve the problems, but they can sure yell and make demands that turn the project into engineering hell. In an incredibly inefficient feedback loop, bad status reports beget closer executive scrutiny of the project, more status meetings, and more overhead and distractions for the engineers. The goal of the closer supervision is to make certain the project is being well run, but a variant of the Heisenberg Uncertainty Principle also applies to projects: The closer you look at a project, the more its direction changes—and the directional change is not usually good.

A coworker managing a difficult project was having a really bad day. When he was called into his third on-demand status presentation of the day, his frustration reached the boiling point. Halfway through the presentation he stopped talking and looked at the handful of executives who had called on him to gain visibility into the project. "Why are you here?" he asked. "Can any of you help me solve the problems that are delaying us? Can any of you give me talented resources that can help solve these problems? Can any of you in any way help me, or do you just want to know what's going on?" The shocked executives were silent for a minute, and then one suggested he take a break and go get a cup of coffee. The moral is that, although the goal of closely observing a delayed project is honorable, it can be very disruptive and stressful if carried to extremes.

The Myth of Managing to a Schedule

In corporate product development one often hears of the need to manage the development project to the schedule. This means that people and events must be organized to continue delivering project milestones at the schedule designated times. The great mystery is how to accomplish this. One popular approach is to hold regular status meetings where individuals responsible for subsystems, components, and outside vendors report on the progress of their areas to the group. The project manager questions those who report they are falling behind schedule about reasons for this and about their plans to get back on schedule. Sometimes these meetings have substantial attendance of senior individuals. It is not uncommon on large projects for a single status meeting to consume several thousands of dollars in salaries. If the project starts falling seriously behind schedule, it is relatively common to address this problem by holding daily status meeting to ensure everyone is well focused on the correct tasks. If things don't get better, adding additional staff or perhaps mandatory overtime may be considered, along with shedding features and shortening time devoted to testing.

It can be beneficial to gather together a reasonable crowd of people making contributions to diverse sections of a project. It can be beneficial to hear about activities of other groups to determine if there is duplication of effort or other inefficiency. However, a large status meeting is the wrong environment to exchange important development information. If something is not working or is poorly understood, the engineer should pick up the phone, send an e-mail, walk over to the other office, call a meeting, or get on an airplane. Engineers that are confused or need to know something should not be waiting for a status meeting to ask questions, and good project managers spend time with individual engineers on a regular basis to certify that there is no confusion or work stoppage due to missing information.

Daily status meetings are way too much overhead to focus the engineers on appropriate topics. Any needed focusing can be accomplished in much smaller groups or even by e-mailed short lists. Daily status meetings are more to convey a sense of urgency to the engineers and for the comfort of members of management, who need to feel they are contributing to the success of the project. Think about it: If moving from weekly to daily status meetings actually helped, moving to hourly status meetings would help even more. Though it is true that those present can identify the most important problems and can set priorities accordingly, there is overhead in interacting with the engineers and interrupting their ongoing activities to accomplish this. There is overhead associated with each context switch. People cannot stop working on one thing and immediately become productive on something else. Constantly switching tasks and priorities leads to thrashing, low morale, and lost productivity. Although everyone will agree that product development is a dynamic environment, to a great extent our management skill will be measured by our vision and anticipation of the future and our ability to keep people focused on a finite number of relevant tasks. Management fails this test if it repeatedly changes directions and priorities or is repeatedly adding and deleting product features.

I was in a status meeting once where two managers were arguing for several minutes about the priority of a bug. A staff member excused himself from the meeting and came back a few minutes later. "Okay," he said, "I fixed the problem." The point is that a great deal of effort goes into talking about and prioritizing and reprioritizing bugs and features to make sure that the most important (or perhaps most catastrophic) problems are worked on first. What is often lost in all this overhead is that such activity only makes much sense if the intention is to ship a product before a bunch of the bugs get fixed. Moreover, the schedule savings achieved by consciously delivering a lower-quality product can be minimal given the amount of time spent talking, meeting, and refocusing the engineering activities. If you really plan to fix most of the bugs the best order would be to fix those likely to expose more problems first. This order may be very different from the order usually used, which is to prioritize fixing the bugs most visible and disruptive to the consumer.

Perhaps development can be accelerated with the addition of more staff? Additional staff would be more likely to help early in the project when there are more things going on simultaneously. Unfortunately, the system integration phase is where many projects begin falling seriously behind schedule. System integration is where the product as a whole first comes together enough to expose oversights and systemic problems. By then much of the independent work that could have benefited from more developers has already been completed. Adding additional resources late in the project is very inefficient. The new individuals must learn about the project development practices, but to be useful fixing problems detected during system integration they must also learn about the detailed interaction of multiple subsystems. Worse, training new people puts a short-term burden on the existing staff. Although it runs contrary to some management theories about maintaining schedules, the ideal project actually sheds developers later in the project. The reason for this is that most work in later stages of a project involve system integration issues and tracking down bugs across multiple subsystems. Like having a number of people digging holes in a confined space, lots of developers making system-wide changes get in each others way. In addition, you want the best people with the broadest system knowledge doing the final optimization and bug fixing. It's not a good idea to add a number of new developers during the later stages of a project.

I was sitting at lunch one day when a person involved in managing a long-delayed project asked my advice for getting the project back on schedule. I responded that half the people on the project needed to be removed from it. My answer so much contradicted the manager's sensibilities that he thought I was insane. In fact, the project in question never actually shipped and died a slow horrible death after several years and many dollars. Numerous people working on a disorganized project is not a plan for success, and adding more is not likely to help. Newcomers at a late stage burdens the existing staff and may take weeks of training before they become useful. However, there is another negative that should not be underestimated. Adding additional people also increases the burden to plan and track their activities. Adding more staff without a comprehensive plan of where to use them, how to train them, and how to track their contribution is clearly not a quality approach to project management.

If not additional staff, can required overtime help? Perhaps, but it depends on many circumstances. To be able to ship a product with any amount of pride, all but the most insignificant bugs must be found and fixed. However, you can't just tell the engineers they have to solve all the problems by some deadline. While I was writing this book a friend of mine got laid off. He was working for a company that bet heavily on winning a government contract. Unfortunately, a demonstration for the government agency revealed numerous bugs. The president of the company told the agency it needed a month to fix the problems. A month later, few of the documented problems were fixed, and the entire system had become unstable. The government agency disliked what they saw and refused to issue the contract. Failure to win the contract forced my friend's company to release more than one third of the staff. In the meeting where the president announced the layoff he blamed the engineers, saying, "We told them what they needed to fix, and they failed us." What the president needed to understand but probably did not learn was that you can't just order engineers to fix complicated problems on a hard deadline. It doesn't work that way.

Required overtime may help a project, but engineers with pride in their work and good morale would already be working long hours. Mandatory overtime sometimes backfires into lower morale and correspondingly lower productivity. It can also result in stupid engineering tricks, where problems are solved in bad ways just to make them go away. Required overtime can be very much a negative if the duration is open-ended or if the goal is irrational. Few will see the sense in requiring overtime to finish a product in a month if knowledgeable technical folks believe the product is a year away from completion.

If you can't keep a project on schedule with more status meetings, more staff, or working longer hours, can you eliminate features or reduce testing? Perhaps, but once again it depends on many circumstances. Eliminating features in later stages of a project is not a panacea to schedule woes because the hard work of creating the infrastructure to support the feature may have already been done. Eliminating a small amount of remaining development or the visible aspects of a feature may not save much time. Reduced testing, however, can dramatically improve the schedule. In fact, the less a product is tested, the fewer bugs will be found. If the goal is to quickly ship something—anything, even a piece of junk—reduced testing is the right answer.

A reasonable case can be made that the ultimate result of attempting to manage to the schedule and to force the product to be delivered on the designated date is a lower-quality and feature-reduced product. Given the generally poor quality of many products on the market you would have to guess that this is a relatively widespread approach. Is management helpless to maintain the specified schedule without sacrificing quality? Is there anything that can be done to keep a project from falling behind? Keeping the project on schedule, especially on a really aggressive schedule, may not be possible if serious problems are encountered. However, there are a few things that shine above all else as far as not making the situation worse. Do not lock the engineers in status meetings for hours every day. Do not give them a daily update to a bug priority list. Do not continually pester them with questions about when problems will be solved. In most cases if they knew what the problem was it would already be solved. What is really important is to do everything in your power to let the engineers actually work on the problems. Can you be creative in getting the engineers to work harder? Probably, but this is a case of being careful what you wish for. You really don't want engineers to work on problems; you want them to solve the problems. There is a big difference. Engineering is not like digging a hole. Sometimes working harder is not the right answer; working smarter is.

This engineering stuff is hard. You have highly intelligent, highly educated, and highly paid people working with determination to overcome entropy and to create a complicated device from a number of components and subsystems. Sometime the universe is incredibly uncooperative and strongly resists attempts to organize its atoms into a cool new product. In the book *The Strange Case of Dr. Jekyll and Mr. Hyde* by Robert Louis Stevenson (1886), Dr. Jekyll becomes Mr. Hyde by drinking a special potion and returns to his Dr. Jekyll persona by drinking a second potion. A critical twist in the plot (I don't think I'm giving away any secrets since the book is more than 100 years old) is that Jekyll runs out of the second potion. When he makes more, it doesn't work reliably, and he becomes progressively more desperate as his analysis fails to find any problems in multiple attempts to recreate the working potion. Finally, he is forced to conclude that some unknown contaminate was responsible for the success of the original batch. This is the prototypical Dr. Jekyll paradox, where an incorrect version of something works but the correct one doesn't.

It is not unusual to find in engineering that a mistake actually allows a bad design or implementation to somewhat work. When that mistake is fixed, everything breaks, and it takes days or weeks to track down and fix the real defects. It is exactly the Dr. Jekyll paradox class of problems that is responsible for the worst and most unexpected delays in a project. Some masking error or event creates the erroneous impression of functionality. Only when that issue is corrected can the depth of the underlying problems become visible. It is unfortunately a hard fact of life that creating new technology cannot always be done on a schedule and that managing to the schedule can only be espoused until complex things start going wrong. People are people, however, and their attention span consists not of how they got here but where they are. The very same executives that insisted on an absurdly aggressive schedule may hold the engineers and project management responsible for delays.

Humans seem to simply have a widespread inability to accept that there are times when they have little or no control over their fate. As a working engineer I have repeatedly seen the damage to workplace morale and the related impairment of product quality that occurs when extreme measures are used in a desperate attempt to keep a troubled project on schedule. The belief that some clever management or motivational technique can be found to rescue a sliding schedule has been proven false on a staggering volume of projects. But like a gambler on a losing streak, the hope is that the right lucky charm or prayer or inspiration will be found and that it can be used to win the next hand and keep the project on schedule.

The Myth of Managing a Vendor

There is no myth involved with managing a vendor providing a standard component. Sometimes bad things happen, but in general managing such a vendor is reasonably straightforward. It involves negotiating a solid contract, staying on top of deliverables, keeping track of details, and having a fair amount of common sense. Things become dicey when the component is actually a new product that the vendor must design, develop, and deliver on a schedule compatible with your product. The vendor is just as likely to be late on its development effort as any other company creating a new product. Sometimes an entire set of schedule dominoes exists as a sequence of progressively more complex products critically hinge on the successful and timely delivery of aggressively scheduled components and subassemblies.

The safe approach is to use only existing and proven components in the development of your new product. It is a fantasy that you can manage a vendor into maintaining its stated development schedule of a new component. It is a fantasy because there is no guarantee that you can manage a complex development effort to an aggressive schedule within your own company. Why then would you assume you can do better at managing a vendor? There are often inspirational corporate statements such as, "We cannot allow the vendor to fail," or "We must not blame any problems on the vendor because it is our job to ensure its success." Though these sound good, they generally amount only to wishful thinking because there are very few real options for helping the vendor succeed. Most vendors carefully guard their proprietary technology and absolutely will not allow external staff to come and assist them no matter how catastrophically behind schedule they are. Penalty clauses in the contract do little if there is no viable alternative to the vendor. Should you reach the point where you decide to invoke penalty clauses, there will likely be legal wrangling as the vendor recites numerous real or fabricated instances in which its work was delayed by your company. Should you win the legal arguments, you may lose anyway. If the delays are great and the penalties are severe, there may no longer be sufficient incentive for the vendor to make the component at all. What do you do if there is no alternative?

Perhaps the most misguided line of reasoning is that your company is a valued customer and that therefore the vendor's management team will put its best people on the component development project. It could be that the vendor really does apply its best people to the project. Unfortunately, even the best may not be able to deliver on an unrealistically aggressive schedule agreed to for the sole purpose of winning the contract. This is a most serious risk when the company suggested the "right" answer by demanding delivery by a specific date instead of simply asking the vendor when the component will be available. A nearly universal marketing strategy is to agree to whatever date is required to win the contact and then to beat the engineers to achieve it. Vendors know that even if they are late, few customers will incur the pain of converting to a different and incompatible component. Customers will vigorously complain about delays, but, once committed, rarely do they cancel the contract and find another vendor.

It is also possible that the vendor's "best people" are in fact the most trusted and most relied on—but not the best—engineers. Sometimes those viewed as the best people are those who rarely dispute executive direction and exude optimism and confidence in the success of the project. Unfortunately, the hopelessly optimistic can-do Pollyanna employees might not be as good at actually delivering the component as the pessimistic and negative Chicken Little engineers. My experience has been that there is often good reason for pessimism even if it is unpopular. Negativism and pessimism are not a self-fulfilling prophecy as might be claimed by upwardly mobile effervescent managers. In reality, negativism and pessimism allow the development team to focus earlier on looming problems and to overcome them more rapidly or even to avoid them altogether.

Having a perpetually optimistic manager leading the development of a needed component may be a particularly bad situation. The customer will receive a constant stream of good news about each component delay. The positive political spin and continuing statements that the problems have been solved make it difficult for the customer to get a true sense of an accurate delivery date. In a worst-case scenario, promised delivery dates will be repeatedly missed as surprise delay after surprise delay accumulate and turn the customer's product development schedule into a disaster.

When all else fails, managing a vendor seems to escalate to vendor brutality. At this point the corporate statements become something along the lines of, "They've been late on their deliverables for a while now, and we're just not going to accept that anymore." Ordering people around and telling them what must be done is not the same as helping them succeed. Simply demanding that a component be completed on a given date and forcing the vendor to agree proves you are a take-charge manager and provides a clear scapegoat for a failed project. Unfortunately, this approach is not often a path to success. Having someone to blame is not the same as achieving success.

Schedule Remediation

The preceding negativism about keeping a project or vendor on schedule is only appropriate when delays are the result of difficult problems—or perhaps an unexpectedly large volume of problems—taking a long time to solve. Schedule delays can readily be minimized or recovered if they are due to inept developers or bad project management. Unfortunately, closer supervision and additional status meetings rarely make developers smarter or project managers more capable. A bad developer can destroy a project in an instant. Even if all the work of bad or weak developers is examined in detail, the developers remain a danger to the success of the project. The burden and risk they introduce is generally unacceptable. The only safe and efficient approach is to remove them. Likewise for bad project managers—no matter how closely supervised, they generate chaos and disorganization by repeatedly focusing resources on inappropriate activities and giving poor guidance and direction. Even if senior management corrects course on a daily basis, the damage done through the turbulence of repeated direction changes adversely affects the project and morale of the team.

If senior management believes that a project has fallen behind schedule due to less than optimal project leadership it is management's responsibility to correct the problem. Telling the current project manager to fix the problems and to get the project back on schedule is not really helping him or her to achieve success. Continuing to provide direction through an intermediary project manager whose performance is lacking rarely improves the situation, even with tighter oversight and more frequent status meetings. It is vastly superior for one of those who believed the project could have been run in a better fashion to replace the current manager and to take direct and personal control of the project. If all goes well, the guiding hand of the more qualified manager will reduce chaos and will precipitate faster and better solutions to problems and get the project back on track. If it happens that problems are not solved fast enough and the project continues to fall further behind schedule, then the process of replacing the project manager must be repeated until someone is found who can succeed or the company concludes that the problems are simply too difficult to be solved in the scheduled time.

The only way to guarantee improved performance of the team is to remove the weak members and too replace them with more capable staff. At some point the company may conclude that its staff is just not smart enough to accomplish the desired goals. The company must then decide whether its goals are irrational or it needs to hire better people. Either way, pressuring, berating, and abusing the existing staff does not improve performance. In fact, it has the specific effect of reducing the quality and functionality of the product as the staff tries desperately to accommodate the schedule demands being made of them.

Is the art and science of project management doomed? Is there no way to coach developers into getting better or to accurately create and adhere to a schedule? Developers, project managers, and everyone everywhere can be coached and can become better at what they do. Unfortunately, coaching and learning are not readily compatible with tight deadlines. Schedules are much more accurate for familiar and well-understood tasks. More correctly, schedules are much more accurate when made by those experienced at performing the specific tasks. New and unfamiliar development is very much at risk since the technology and associated difficulties may be new to the available staff. In extreme situations, such as building an antigravity device, nobody on the planet (at least none I'm familiar with) knows enough to accurately predict the duration of the project. Any such schedule is nothing more than a wild guess and the subtle workings of the universe cannot always be discovered on a fixed schedule. A real-world project need not be an antigravity device to create unexpected problems for unfamiliar developers. The worst-case scenario occurs when those responsible for creating and managing the schedule don't know they lack needed knowledge and ignore the pessimistic and negative folks urging caution.

When delays are caused by serious problems and not bad staff members, it seems there is little hope of preventing complex and aggressively scheduled projects from falling behind. Is there any hope of recovering lost time and getting back on schedule? Consider what this actually means. To get back on schedule means that the remainder of the project goes better than expected. However, most schedules are made to anticipate success. There is no planned "hold" in the countdown that can accommodate lost time. For all the reasons previously considered, tasks are optimistically scheduled with minimal buffers and pads. Given this situation, the only rational approach to recovering lost schedule is to reduce the amount of work or to increase the speed with which the work gets done. As discussed previously, both of these tactics have inherent limitations and consequences and cannot guarantee success in maintaining the schedule, much less recovering lost time.

Powerful personalities may attempt to reclaim lost time by sheer force of will. Assertive and dominant corporate leaders may find schedule slips simply unacceptable and may go to extreme measures to assert reality. This occurs when a pronouncement is made with such conviction and belief that the fabric of reality itself is bent to the will of the individual. Although I have never personally seen schedules recovered with this approach, common sense dictates that it must be possible or intelligent and educated corporate leaders would not use this schedule remediation method. Unfortunately, my direct experience has been that executing extreme measures in a vain attempt to bend reality causes project chaos and stupid engineering tricks and seriously damages staff morale. All of this leads to increased project delays instead of the intended schedule remediation.

Working harder, adding staff, or eliminating features are classic methods of maintaining or recovering schedule but have a relatively poor success rate on difficult projects. Occasionally, a drastic course correction is necessary. One successful approach to schedule recovery is to rename the project and to create a new schedule. The old and catastrophically delayed project slides peacefully into forgotten history while the new project takes front stage with a much more realistic schedule. If all goes well, success, praise, and career advancement awaits those whose skill and vision brings the new project to timely completion.

Schedule Revisions

For some projects, the time comes when delays can no longer be hidden and all sane people have given up hope of getting back on the original schedule. Once the decision is made to rework the timetable, one must still decide on the actual goals and responsibilities of this effort. There is a fear that technical folks will sandbag and ridiculously pad a schedule, so who will be assigned the task of creating the new one? Should this job fall to the same folks who produce the first schedule that was so inaccurate it had to be abandoned? Must original assumptions be changed? Must original goals be changed? Arriving at the decision that a new schedule is needed is traumatic for an organization, but the challenges go far beyond simply writing down a new to-do list. Having been wrong about the initial schedule is an opportunity to learn from these mistakes with the goal of making an accurate schedule not only for the current project but also on all future projects.

Corporate executives like to encourage positive and optimistic people, but a project schedule shouldn't say what might happen or could happen or describe the best-case scenario. A project schedule should say what is likely to happen. Resources cannot be properly allocated, and tasks cannot be properly sequenced if hopelessly optimistic people continue to describe project scenarios that fail to materialize. It really, really helps if the project manager and corporate executives have a sufficient understanding of the technology and the project to understand a reasonable development time. The support of management allows correct setting of expectations and proper and timely coordination of the project effort across all divisions of the corporation.

One may claim that currently used project development methodologies must change. From cell phones to televisions and toaster ovens, products have gotten incredibly complicated. Even a casual observer can see that the quality of many modern products is not very good. Perhaps a paradigm shift is needed. Rather than performing extreme measures or exerting destructive pressure in a vain attempt to manage a project to a hopeless schedule, it may be better to periodically reproject a timetable as new information becomes available. Far from a costly or defeatist approach, this new paradigm removes any incentive for status misdirection and enables those able to foresee upcoming problems to make use of that skill. Regular reprojections also allow realignment of resources and minimize waste from unexpected delays. More importantly, such an approach has a great possibility of actually resulting in a shorter overall project schedule since resources are regularly realigned to delaying problems and engineers are not distracted by burdensome status meetings and direction changes.

One of the most successful projects I was ever involved with suffered an extensive delay at the end of the project due to a detailed and convoluted third-party certification process. This delay gave the engineers time to "tweak" the implementation and to clean up piles of "minor" bugs that were known but not fixed due to schedule pressure. The resulting quality, performance, and superior functionality allowed the product to overtake and surpass the established market leader.

Purposefully delaying a product and continuing to hone it after it is "done" is actually a financially sound concept. The additional cleanup minimizes field failures, returns, and customer dissatisfaction. Shipping a superior quality product decreases the continuing manpower drain associated with addressing problems from a bad product already in the hands of customers. Unfortunately, most corporate bonus plans, reporting hierarchies, and annual appraisals are not conducive to such conservative and dawdling behavior. Engineers seem to be doomed to a never ending spiral of aggressive schedules and late delivery of buggy and feature reduced products.

Software Update

Many corporate cultures very much support and endorse optimistic behavior by managers and staff. Software update is a godsend for these cultures. It allows shipment of a barely working product with the idea of fixing it later. In many cases, software update is the ultimate pay-me-later scenario where the corporation borrows against its future engineering productivity to get a bad product out the door faster. Not only does the corporate image suffer as customers encounter problems, but new product development is also slowed while a team of engineers continues to work on solutions to problems with an already delivered product. Worse, it is always harder to fix problems in an unobtrusive fashion on a product that is being used by customers. The extra care and caution that must be exercised is compounded by more complicated testing scenarios. The software update must be tested with a variety of product usage scenarios to ensure the new software can be received and activated without undue disruption of normal product operation. In general, any problems being fixed by software update could have been fixed with much less labor if done before the product was shipped to customers.

In the old days products mostly had to work before they were shipped. Software update dissociates the need for a working product from the declaration of project complete. Software update may even be able to address hardware problems by downloading new software that allows such problems to be hidden or minimized. Software update has evolved from the plan of last resort to the intentional method of generating revenue from a product while development of that product continues. Unfortunately, development of new features can be greatly constrained and made much more difficult by the existing functionality already in the hands of customers. Any new features or changes to enable new capabilities must concede and accommodate existing customer usage. Normal development delays are compounded by these constraints, and the development of new features for software upgrade takes much longer than if the features were provided with the initial delivery of the product.

Formal Bug Tracking and Metrics

Engineering projects go through several ordinary and very predictable stages of maturation. The projects may have different names, features, and applications, but after you've seen a few they begin to look remarkably similar. The first time the low-level driver software is put on newly developed hardware, nothing works and the project appears doomed. Two or three days later, some of the board is working; after a couple of weeks the device may have a few quirks, but it is generally usable for subsystem development. For some period of time, depending on the complexity of the product, individual subsystems are developed and unit tested by the engineering staff. During these early stages so many things are wrong and the status is so dynamic it is generally pointless to formally track bugs. Any such activity would be extremely counterproductive since the engineers are fixing numerous personal bugs and would spend far more time documenting the bugs than it would take to actually fix them. At some point the subsystems are working well enough to move the project into the system integration phase. The first time the critical subsystems are jointed together the project again appears doomed as catastrophic failures occur for unknown reasons. Although things seem very bleak, these early integration problems are usually just stupid mistakes and oversights, and soon the platform is able to lurch and limp and perform some of the expected operations.

System integration nearly always takes longer than expected because engineers are human and make bad assumptions, read the requirements wrong, or flat-out screw up. After a few days to weeks, the most blatant of the errors have been corrected, and the platform starts to show signs of life. At some point it becomes time to start formally tracking bugs, their priority, and their resolution. The decision point as to when this should happen is a mixture of practicality, corporate culture, and the personalities at play. There is the continual desire of management to have more visibility into the progress of the project and the continual reluctance of engineers to spend time on such "overhead" activities. Entering a bug will be mechanically straightforward and efficient if an appropriate bug tracking tool is being used. Mechanical ease does not obviate the need for the person entering the bug to think and adequately describe the bug and the conditions under which it appears. It is reasonable to expect that the typical bug takes several minutes to enter and several for someone to read and understand.

It always seemed very reasonable that logging a personal bug was silly and counterproductive if it took less time to actually fix the bug. After all, the point of bug tracking is to improve the quality of the product and the speed and efficiency of project execution. Requiring entry and tracking of every bug burdens the project with additional and arguably unneeded work. Bug tracking has the greatest value when used by customer service and testing organizations to log bugs for the engineers to fix or to provide a forum for tracking progress on complex and long-duration bugs. Not everyone agrees with this viewpoint. Sometimes the business of bug tracking and metric gathering takes on a life of its own. Corporations have entire groups, bureaucratic hierarchies, and numerous jobs associated with the process and ceremony of gathering, manipulating, and distributing a broad variety of metrics. There are groups and cultures where metrics themselves are the job and not merely an activity to support product development. The metrics may include the number of bugs discovered and fixed, the number of lines of code written per day, the number of hours of hardware design per square inch of board, the number of customer returns, and a generally diverse set of numbers to describe every imaginable situation. There may be battles and skirmishes between engineers and project managers, who view metrics as a way to improve the development effort, and the "metric Nazis," who gather and correlate numbers so they can do presentations.

Some metrics are indispensable for developing, manufacturing, and maintaining a quality product. Prior to delivery a trend in the number of reported but not-yet-fixed bugs provides an indication of the true state of the project. The product is nowhere near ready to ship if the number of unresolved bugs is increasing from week to week. It is a correspondingly clear sign that development is approaching completion if the number of unresolved bugs is declining. Tracking reasons for warranty repairs and returns can quickly identify problems with components or assembly techniques once a product is shipping. It is also useful to have an accurate and comprehensive bug database that can be checked to see if similar symptoms have been reported or to find documentation of temporary workarounds.

Bug reports are most useful if they are logged with meaningful names and accurate descriptions of the problem, how to reproduce it, and how to avoid it. It can also be worthwhile to indicate whether the problem was introduced during the design or implementation stage of the product and the versions of hardware or software to which the bug applies. A great deal of information can be captured about every problem and quirk, and a number of the metrics tools allow various fields in the log to be marked "required" to ensure certain data are entered when the bug report is made. Gathering all this information is critical to the industry of bug tracking and the lifestyles and corporate importance of the metric Nazis. Unfortunately, gathering all this information is not free. It takes time to enter the data, time to process and prioritize them, time to graph and chart them, time to present them, and time to attend the presentation. There is not only the direct cost of all the work but also the lost opportunity cost; that is, what beneficial things were not done because of the work associated with the metric bureaucracy?

Over time, profitable corporations tend to trim money-losing activities and expand money-making ones. It would seem that corporations with extensive metric bureaucracies must have found cost benefit in that activity. Some have, but there are also organizations that spend the money to gather the data and use them exclusively for presentations. Never have they used the metrics to improve personnel or process. In some corporations an absolutely remarkable amount of effort, and therefore cost, occurs for the sole purpose of keeping curious executives informed. A number of corporate managers feel the need to be involved in projects even if such involvement only amounts to attending regular presentations and complaining about delays. Few corporate leaders have the time or ability to make genuine contributions to a technical development effort, but the premise of the need for project visibility is engrained in the ethos of many corporations. Seeing project metric presentations and emphasizing the need to minimize negative trends and to encourage positive ones allows executives to feel like they are facilitating a project and leading the company.

Good metrics, when used well, have the potential to substantially improve company operations. This is documented and proven by numerous studies. The problem is that some corporations consciously or unconsciously choose to not learn from the metrics that were amassed at considerable expense. This may be from laziness or poor execution but occasionally is because the metrics don't present the desired or politically correct results. I once sat through a metrics presentation from a corporate quality "tiger team," which had determined the reason for a large number of field failures and customer returns to be the result of a small number of high profile bugs. The presentation called on the engineering group to improve its process to prevent shipping future products with similar serious problems. Unfortunately, to the embarrassment of everyone involved, the bugs were already logged in the engineering database. The called for engineering processes were already in place. The problem was that a senior executive in the business group had decided that product delays could no longer be tolerated and the product had to be shipped, warts and all. In another situation, analysis of historic product metrics produced the surprising result that superior product quality was associated with a small number of engineers. Amazingly, the identified engineers did not have a high corporate profile, and some were even disliked by management. This analysis effort ended with little fanfare and no recommendations for improvement.

From a purely engineering perspective it is desirable to maintain a conveniently accessible database of bugs of significance. In the best possible scenario, hardware bugs, software bugs, mechanical issues, manufacturing issues, and every other contributor to the product maintain a meaningful list of bugs in a communal database. Logging trivial bugs or hiding problems adds overhead and delays resolution of important issues. This potentially large list of bugs needs to be periodically reviewed, or "scrubbed," to ensure that responsibility is correctly assigned and appropriate progress is being made.

There is a fairly clear distinction between two classes of people that attend bug scrubs and status meetings. One group consists of people with the depth and breadth of knowledge to make contributions to successfully building a quality product in a timely fashion. These folks have the knowledge to understand the significance and complexity of bugs and delays being discussed and are in a position to offer useful suggestions to resolve the difficulties. Their involvement can help the meeting be a productive session that contributes to successful and rapid delivery of the product.

The second class of people attends project meetings to satisfy their curiosity about the progress of the project. This curiosity can be variously termed a "need to know" or a "need for visibility," but regardless of the phrasing these people are simply seeking project status and are incapable of making a bona fide contribution to the process of developing and shipping the product. At times these individuals may actually damage progress on the project by focusing on insignificant or irrelevant topics or trivializing issues they don't understand. Sometimes this is an extremely difficult situation for those involved in the project because the individuals are executives in the corporation whose suboptimal direction must be followed.

From a pure project efficiency perspective, the smaller the number of people involved in the bug scrubs the better. Adding more people exponentially increases the time involved as each person comments, or perhaps postures, on various problems. In fact, the project manager privately reviewing the list and meeting separately with individual assigned engineers is incredibly cost-effective. Big status meetings have a purpose, but they are far less efficient than one-on-one interaction of a small number of people focused on fixing problems and shipping the product. Corporate leaders must stay informed about progress on numerous initiatives to do their jobs, but gathering data and creating presentations toward that end can be a distraction with respect to actual product development. Excessive bug tracking, metric gathering, and status seeking can move beyond distraction and can become burdensome. Caution must be exercised to ensure such expenditure of time and money yields productive benefit beyond giving a curious bureaucrat visibility into the project.

Tracking and sharing information on bugs associated with a vendor-supplied component provides a special challenge. In this situation, the engineering need for information combines with the management need for visibility and encounters a corporation's natural resistance to showing its dirty laundry. Secrecy within the same corporation may be hard to understand and accept. How, for example, could one condone the hardware group keeping functional problems secret from their sibling software group? However, allowances must be made for incomplete sharing of bugs between corporations.

Public disclosure of certain bugs may put a vendor at a competitive disadvantage. This becomes especially delicate when the company must work around quirks or bugs in a vendor-supplied component. The more challenging component problems are those that occur only when the component is used in the unique application of the product itself. These are especially tough, and resolution requires close coordination between the involved companies.

A common bug database between companies enables superior collaboration and sharing of information about bugs and the mutual setting of priorities. Care must be taken to ensure that process and procedures are in place to only expose needed information and to maintain the secrecy of proprietary data. In this scenario the bug-tracking tool must support the concept of rights management. Corporate "bug privacy" can be achieved by restricting the right to view bugs to authorized individuals only. From a practicality perspective, it is also beneficial if the vendor contract requires prompt disclosure of relevant component bugs. This avoids the situation where the company's engineers have to track down problems and prove the component is at fault before the vendor takes action on an already known problem.

The project manager has the difficult job of herding the developers into compliance with accepted project management practices and convincing them of the need and benefit of detailed metrics and of the need and benefit of management visibility into the project. The project manager must be able to respond to difficult questions from diligent and intelligent people and to adjust the contrarian viewpoints of some working on the project.

Cynical engineers may claim that those involved in the effort have a pretty good idea what is wrong with the product. As such, they may claim the labor and cost of zealous gathering of metrics and regular and detailed management presentations of this information does little to benefit them or expedite the project. Cynical engineers may further claim that the widespread practice of fancifully exaggerating the positive aspects of project status and minimization of problems reduces the already marginal value of these grandiose management presentations. Cynical engineers working hard to solve complex problems may think that these contrived presentations, produced and attended at significant cost, are solely for the edification of curious executives who then complain about project difficulties and delays but are incapable of helping the engineers overcome them.

One must feel the pain of the project managers trapped in such situations. On one hand, there is some merit to these contrarian views. The project visibility gained by executives through the prism of contrived presentations and doctored demonstrations certainly reduces the value of the time and money that went into them. On the other hand, corporate leaders absolutely need information, metrics, presentations, and demos to do their jobs. One indeed must feel sympathy for project managers trapped between these two worlds.

Formal Testing

Testing in various forms has been happening since the start of the project. At first there was testing of ideas by discussing them with others. Then there was testing of modules and subsystems by individual engineers. As more pieces of the project are pulled together, ad hoc system testing begins to explore progressively more complex usage scenarios. Ultimately, it is time to move into formal testing of the product. As with formal bug tracking, practicality, corporate culture, and individual personalities determine when this actually happens.

Formal testing depends on test plans written or adapted specifically for the unique product being developed and perhaps even a specific application of the product. These test plans often strive for maximum coverage of both the visible and internal workings of the device. The test plans spell out realistic and contrived situations to stress and validate as much of the design and implementation of the product as possible. Writing test plans can be a collaborative effort between multiple corporate organizations. Engineers with knowledge of the implementation define scenarios that exercise specific blocks of hardware or sections of software. Business and marketing groups can provide tests to ensure that the user interface is correct and the product performance is suitable. Testing specialists can add their expertise to define equipment configurations and methods of automating tests and collecting results.

Appropriate test plans must cover the entire spectrum of product operation, from initial installation to normal operation and even to ensuring graceful degradation under adverse conditions. Each plan must be executed and the results observed and documented. When the product fails or behaves in an unexpected fashion, a bug should be logged in the bug-tracking system. Subsequent versions of hardware and software must be retested to verify nothing previously working was accidentally broken when the new version was made.

Testing professionals need a reasonably pessimistic and meticulous mindset that is not overly accommodating to the engineers. I was once in the lab discussing some issues with the manager of the test team. At another workbench a staff member could be seen running product tests. As our discussion continued, the test being run by the employee was visibly failing. Once, twice, three times it failed. On the fourth try, the test passed, and the person could be seen making a notation in the testing logbook. "Excuse me," said the manager, as he walked over to the other workbench and observed that the test was marked "passed" in the logbook. The manager remarked that he had seen the test fail several times. "Oh yes," said the tester proudly, "I had to run it a bunch of times to get it to pass."

As the previous story illustrates, there are good testing methodologies and bad ones. Unfortunately, most people make rather poor testers by innate personality and must be coached into the proper mindset. By nature, most people are far too accepting of what constitutes reliable performance. When something "works," it works every time. An astrological horoscope doesn't work if it said you would have good luck on the day you wrecked your car. A system for beating Las Vegas or picking stocks doesn't work when you end up penniless. The reason systems and horoscopes and other problematic methods of divination still exist in the world is because a great many people have remarkably low success criteria. This is an unacceptable trait for testers of electronic products, especially when such devices exhibit statistical problems.

Statistical bugs are those that don't always occur. For example, a product may sometimes fail to respond when powered on. Often problems like this are first sporadically reported by random people. When the test team looks into the problem it cannot be duplicated and the product seems to work. Over a period of time anecdotal evidence builds into factual evidence, and eventually the product team must accept that such a problem exists. Once this is known, the first step is to characterize the problem. The test team may turn several products on and off hundreds of times at different temperatures, different humidities, and under varying circumstances trying to identify a pattern to the failure. Sometimes a pattern can be found; other times the problem simply randomly occurs some percentage of the time. The important point is that seeing a test work once doesn't prove there is no problem or that a previously reported bug has been fixed.

It must be emphasized that formal testing is an attempt not to force quality into a product but to validate that such quality is already present. Formal testing is no substitute for competent and conscientious design and implementation and cannot be used as a last-minute Band-Aid. Sometimes formal testing identifies characteristics indicative of poor engineering. Memory leaks, for example, are the result of allocating system memory but not returning it when the operation completes. Memory leaks are not rocket science. They occur because somebody was not conscientious in his or her management of memory allocation. Most often this happens when a process has multiple exit paths and only cleans up memory on some of them. Everybody screws up memory management from time to time, but numerous memory leaks in a product may be indicative of novice or careless programming. Observant testers can detect a broad variety of repetitive bugs in addition to memory leaks. Such informative trends should be discussed with engineering management to determine if a general cleanup is warranted rather than gradually finding and fixing a sequence of individual bugs.

Laziness, cost, or schedule pressures occasionally result in the phenomenon of ornamental engineering. This is when the real problem is not fixed because it is easier, cheaper, or faster to compensate for it in some other area of the product. Compensating for a bug instead of fixing it can have unexpected negative consequences. Perhaps the compensation itself results in secondary errors, or perhaps the compensation was not applied everywhere it was needed. Formal testing may detect a pattern of bug migration as patch after patch is applied to correct bugs generated as the effects of the original bug compensation ripple through subsystem after subsystem. Such patterns should be brought to the attention of project and engineering management so the root problem may be addressed.

Finally, managers of the formal testing activity can find themselves in the extremely unenviable position of having to deliver, in the face of mounting pressure, the bad news that the product is not yet ready to ship. Formal testing may be viewed as the final opportunity to prevent a bad product from getting into customer hands. As such, the leaders of formal testing must have the fortitude to tell important people that the quality of the product is not yet acceptable.

Manufacturing

The product has been designed, implemented, and tested; now it is time to start building it. Long before the production line starts running, fundamental decisions must be made with respect to the philosophy and intensity of the manufacturing testing. This type of testing is distinct from the formal testing discussed previously. The latter exists to verify that the product was designed well and implemented correctly; the former exists to assure the product was assembled properly from good components. It is not an attempt to revalidate the design as every board comes off the production line.

Better and more extensive testing takes longer and therefore costs more. Engineers and testing specialists implement the testing methodology, but the corporate business group decides how much money it wants to spend on testing and therefore decides how reliable a product it wants to sell. This decision is based on many factors including the criticality of operation and the consequences and cost of failure. A company primarily interested in being the low-cost leader may choose the minimum possible level of testing. Correspondingly, a company with a reputation of making superior quality products may be willing to spend a little more to maintain that reputation. It may rationalize this additional expense by using its good name to charge more for the final product.

An incredibly broad range of testing options—and therefore testing costs—exists. A manufacturer may choose to prescreen 100% of the components used in making the product. It may also choose to prescreen only high-risk components or may do no component screening whatsoever. A variety of manual and automated visual inspection techniques exist to check component placement and to look for damage or solder problems. Thorough testing of electronics can approach 100% validation of the entire circuit board. Doing such detailed and complex testing often necessitates invasive procedures and the creation of custom in-circuit test (ICT) fixtures and special manufacturing diagnostic commands. The manufacturing diagnostics put subsystems and components into unusual configurations and modes not used in normal operation. Creating these special commands nearly always requires the expert knowledge of the original design engineers and cannot be done independently by the testing group. The needed collaboration between the two groups can take months and can therefore be quite expensive. Sometimes in-depth product testing also requires access to special diagnostic modes of various components. In these cases the needed collaboration extends to the engineering groups of the component vendors.

Manufacturers might also power up and use every product or some statistical sampling of the products coming off the production line. Infant mortality is a common failure mode of electronics products. Detailed charts and tables exist that describe the probability of infant-mortality-type failures of a given product design. Using these charts, infant mortality can be reduced to an arbitrarily low level by "burning in" the product, which takes time and therefore adds cost to the manufacturing effort. Here again, the corporate business group decides how much money it is willing to spend to make a dependable product.

Initial manufacturing rarely goes smoothly. The first few weeks building a product are likely to produce far fewer working devices than desired. The manufacturing yield gradually improves, however, as hard work, skill, and experience eventually track down the problems and beat them into submission. The long wait is finally over, and success can be declared. With a great deal of pride the team can watch its babies be loaded into trucks and shipped to customers.

11

Epilogue

Introduction

It's been a long and difficult adventure, but the product has finally shipped and the production line is up and running. The distribution channels are being filled, and the first product reviews are surfacing. If engineering has done its job well, success of the product is now in the hands of marketing and sales.

Early Adopters and the Competitor Boost

Many groundbreaking new products experience sluggish initial sales as a new market is slowly created by early adopters. Early adopters are universally loved by technology companies. They don't care that they have lived their lives until now without this type of product. They need one. Early adopters don't care that a better device will be cheaper in a year. They must have one now. The best of all early adopters are those who tolerate bugs and quirks. They need the product so badly that they pay top dollar for marginally working devices of questionable usefulness just to be the first on their block to have one. We love these guys.

Slightly less pioneering products may hit the market with several competing devices already available, and yet other competing devices may be in the process of receiving final touches before being released. Multiple companies making or hoping to make a similar product can actually provide a short-term increase in sales through a phenomenon known as the competitor boost. Despite the Digital Millennium Copyright Act, it is relatively common for companies to buy and dissect competing products. In certain situations—perhaps if the product has been long promised or is in an exciting new market—it is possible an appreciable percentage of the early sales are actually competitors buying platforms to study. In some rare cases this competitor boost may overshadow normal customer purchases and precipitate confusing sales reports.

When competitors analyze a new product they look for identifiable major components, estimate circuit board complexity, and try to determine the production cost of the product. They also look for new ideas and places where the design may depart from standard practice. The general idea is to gauge where the competition is with respect to shipping products and internal proprietary designs. It is also useful to see which vendors are providing components for competing products and perhaps to use the information in ongoing component selection procedures or in negotiating future purchases.

When You Guess Wrong

There are times when the excitement of shipping a new product turns to disappointment and despair. The manufacturing line is running full speed to meet the expected sales volumes, but the anticipated sales are not materializing. Warehouses begin filling as meeting after meeting is called to address the issue. Soon, worry expands to include the vendors, as large volumes of components used to manufacture the product continue to arrive. Warehouse space costs money, components cost money, and product inventory costs money. Something must be done to stop the bleeding.

I've had the misfortune to experience such corporate trauma several times. On one occasion the head of product marketing was asked his plan to empty the warehouses overflowing with product. He had a very concise one-word answer: "arson." In another situation, my company decided to save the warehouse costs of storing products by having them shipped back to the development offices. Several times per week employees would meet shipping trucks in the parking lot and carry stacks of product packages into their offices for storage.

Only so much can be done to save money if marketing and sales can't create some magic that sells the product. Eventually, engineering must work with corporate legal and purchasing to slow or stop the product pipeline. The corporate legal and purchasing groups are needed because contracts were signed with vendors' promising to buy specific quantities of components. Often the price being paid was contingent on this promised volume, and reducing the volume could lead to the vendor trying to charge a higher unit price. Worse, there are cases where the vendor made a custom version of the component and cannot sell the no-longer-needed inventory to any other company. In this circumstance, the vendor may feel, perhaps rightly so, that the company should pay a significant penalty.

Many things can go wrong with the voodoo of predicting sales volumes and market penetration. Perhaps a competitor introduced a dynamic new product or the national economy suffers a downturn. Regardless of the reasons, a failed product takes a heavy toll on the morale of the team that produced it. After working hard and sitting through countless status reviews and bug scrubs, it really hurts to carry stacks of products and components to the dumpster.

Supporting a Successful Product

As devastating as it is to have worked on a disappointing product, it can be far more stressful for the team to have a successful one. The stress results from the need to keep production of the product flowing smoothly. If a product is unexpectedly successful, the team may even need to increase the manufacturing volume beyond anticipated levels. This may necessitate the qualification of new sources of components, the negotiation of new contracts, or the addition of manufacturing and testing capacity. Unexpected success also stresses the customer service and warranty repair organizations. New staff may need to be added, and that necessitates additional unplanned training and more office space.

A successful product means that every disruption in manufacturing is critical. The team must be on constant alert to not let the company down. It must investigate reasons for product failures and ways to improve manufacturing yield and efficiency. It must be continually watchful for component shortage and obsolescence issues that could halt the production line. A successful product means a neverending stream of more work, more component testing and qualification, more price negotiations, and more contracts. But, as the saying goes, these are good problems.

Product Postmortems

Success or failure, disappointment or stressful exuberance, individuals and the team as a whole learned a great deal during the project. Merging these lessons into the corporate culture benefits subsequent projects by providing an environment and a background of traditions with which to judge activities and processes and to anticipate whether they will be an asset or a detriment to projects. Project postmortems can, over time, enhance the corporate culture to provide a supportive atmosphere for successful and timely development efforts.

It should be possible to reach a consensus about what was done right and what was done wrong on a project. The time spent in a postmortem is an investment in improving future projects just as the time spent in a design review is an investment in the quality and speed of delivery of a current project. The shared learning of project postmortems can help improve the knowledge of project managers, corporate executives, and working engineers. On many projects the managers can look back and see things that could have been done better. The marketing and sales groups can understand errors in estimating product demand, can hone their ability to create viable and stable requirements documents, and can improve their performance in many other ways. The engineers, now armed with hindsight, can improve their schedule estimation techniques, can better avoid technical errors, can identify inefficient processes that wasted time, and can more clearly see risk areas.

The key to enhancing the corporate culture is a final project review that distributes the lessons learned. One need not deliberate the things done well on a project. Perhaps a little reinforcement is in order to ensure that the corporate culture and natural instincts of the team continue these positive activities. More benefit is gained from focusing on things that went wrong or could have been done better or more efficiently. Negativism and pessimism are, well, negative and pessimistic, but addressing the things that went wrong on a project can minimize the chances of making the same mistakes in the future.

The first step toward having meaningful postmortems is to ensure the conclusions reached are accurate and not politically motivated. Perhaps individuals can be privately interviewed to distill a short list of the most important things that could have been done better. That list could then be circulated to the project team for review and comment. Finally, a short wrap-up meeting could be held to make certain all the conclusions are agreed on and are accurate and perhaps to assign a few actions to guarantee the corporate culture benefits from the knowledge gained.

Success in improving behaviors and the corporate culture ultimately depends on wide acceptance of the issues and removing the stigma of criticism. As with engineering design reviews, the goal of postmortems must be understood as boosting the efficiency and performance of people and processes and are in no way intended to humiliate or demean an individual. No one should be embarrassed to hear in a constructive fashion, "You could have done better." Indeed, embarrassment should come from not doing better on the next project.

Unfortunately, some personalities in some environments interpret any negative comments about their performance as personal attacks. They deflect the suggestions as counterproductive finger-pointing instead of accepting the constructive criticism and learning from it. Such individuals and the cultures they represent are stuck indefinitely in the same defensive behaviors instead of benefiting from mistakes made on the project. In effect, they continue to replicate the same ineffective or suboptimal actions and attitudes. Corporate cultures that do not intervene only reinforce the acceptability of such obstinacy.

So long as those in question believe that their future promotions and bonuses necessitate a spotless and criticism free record, it will impossible for them to accept that improvement is needed or desired. They will continue to deflect blame and will fail to understand that the issues are not malicious finger-pointing but a call to accountability and adaptation. Defusing the politics is paramount to teaching those with strong political aspirations to learn from their mistakes.

Lessons Learned, Finger-Pointing, and Accountability

As an engineer I once turned in a status report where I said I was falling behind schedule. The report made its way to a few senior people and seemed to contradict the information being provided by the project manager. The project manager sought me out and explained, "You are not behind schedule unless I say you are behind schedule." This made no sense to me and directly conflicted with my perception of reality. However, there is more to project reality than merely technical facts. What I didn't understand at the time was that executives oversee behind-schedule projects very differently from projects that are on schedule. What the project manager understood that I didn't was very simple and was very much based on facts of the real world. If our project were behind schedule it would come under increased scrutiny and would have additional status meetings and "help." At this point, the project gets pretty much off track. The well-intentioned scrutiny by the executives adds overhead and anxiety to the project. Engineers transition from worrying about technical problems to worrying about what the big boss thinks. The project manager had learned this lesson on previous projects and was not about to again make the mistake of saying his project was behind schedule.

A different project manager also wanted to avoid the stigma of being behind schedule so shortly before the official "Code Complete" date he had the entire software development staff enter bugs in the bug-tracking system. In less than a week more that 600 bugs were entered, some simply stating that an entire feature was missing. The official project position was that, although we had a few bugs, we had successfully reached Code Complete on the scheduled date. With Code Complete officially, if not in reality, achieved, I was worried that resources would be pulled off our project and reassigned to another. I was pleasantly surprised to learn that we were not only congratulated for completing development on schedule but were also asked if we needed more resources to address the large number of bugs.

I can't really claim that the two preceding examples are of quality lessons learned, but they certainly represent learning on a project. One of the more important things about life may be to learn from mistakes. Indeed, Albert Einstein defined insanity as doing the same thing over and over again and expecting different results. If we don't learn from mistakes and adapt our behavior, we are doomed to repeat the same failures. Given the obvious value, there is a surprising amount of resistance to doing project postmortems. In my experience, a great deal of this resistances centers directly on one or more "rising stars" in the corporate organization having to admit they or someone in their organization made a mistake or could have performed better.

Rarely does resistance to postmortems arise from engineers. They tend to be intelligent and have educated opinions about things that went wrong. In many cases, they don't fear criticism that may result from the postmortems; instead, they may feel blame was incorrectly assigned to them during the project and hope to set the record straight. Less timid engineers very much look forward to the opportunity to express their opinions in a meaningful forum. More correctly, the more outspoken engineers welcome a chance to correct things that made their job more difficult. Resistance to this type of review most often comes from managers and executives who think the postmortem would be a waste of time, unlike the daily status meetings that occurred for the last several months of the project.

Many factors are at play in the psyches of those who resist an open forum consideration of things that went wrong on a project. Right or wrong, many managers and executives believe they have nothing to learn from such an event, especially if negative comments originate from underlings or peers competing for the next promotion. Certain successful people, or perhaps those who desperately wish to become successful, simply cannot abide to have made a mistake. Sometimes this is an internal issue where they are psychologically unable to accept that they could have been wrong or could have done better. More often this is an external issue where any tarnish whatsoever applied to their corporate image is absolutely unacceptable.

It can be quite difficult to understand the goals and motivation of people who think differently from us. If you happen to think like an engineering geek who primarily wants to play with the toys, it can be hard to understand people who are egocentric, scheming, and driven to succeed at anybody's expense. Although such a conniving thought process may be very unfamiliar and confusing, it is nonetheless real. There really are people in the corporate world who by nature hate and scheme. Externally, some of them can be very personable and charming, but they can be very dangerous because the corporation is not a democracy. As engineers, having the knowledge that a particular individual is a detriment to their project does no good because the engineers' votes do not count. The only votes that matter are those of the individual's superiors. Because higher-level managers and executives rarely walk around and chat with working engineers, there is little or no opportunity for the superiors to learn of the individual's incompetence or disruptive scheming. Project postmortems present an incredible danger to the schemers because discussing project problems in a public forum risks exposing the individual's performance to people whose votes actually matter.

Open discussion of problems shines a light on bad assumptions and poor leadership. Open discussion is a catastrophe for those who wish not to be challenged and simply want their orders followed. Corporate politics, however, lack the federal government's ability to block debate and discourse of a topic by declaring it a national security issue and clamping on a lid of secrecy. The corporate schemers must use other techniques to squelch sincere consideration of potentially damaging issues, and they often block such debate with words and arguments intended to manipulate corporate opinion. As such, people who say, "We don't tolerate finger-pointing around here," may very well be worried that the finger is going to point at them or someone in their organization.

Superficially it is a good idea to ban finger-pointing; however, this is really a knee-jerk reaction to the negative word (*finger-pointing*) used in psychological manipulation of the audience. A much more positive reaction is achieved if one replaces *finger-pointing* with *holding people accountable*. An audience is much more receptive to holding people accountable for their actions and directives. However, holding people accountable is itself somewhat more aggressive than what is really needed in a project postmortem. The idea of a final review of what went wrong on a project is neither finger-pointing nor holding people accountable. It is simply to enable continuing improvement of the process of running a development project. The fact is that errors are made on every project, but they are made by fallible and imperfect humans. The finger must be pointed at individuals—not to penalize them for being human but to expose the error and to learn how to avoid similar ones in the future.

Let's Do It Again—and Again

There is no rest for the weary. If you are working for a successful and viable company you were already working on the next product before you were completely finished with this one. If the product was well designed and well

made there should be minimal distraction as you move into working on the new one full time. However, if your company was overly aggressive in sending the product to market, you will likely be analyzing field failures and preparing a software download to fix longstanding and recently discovered bugs. If your company has had a long history of shipping products before they are quite ready, you may be working on two or three or more earlier products as you struggle to find time to begin work on the newest one. And when that newest one is complete and shipping, you'll have yet another one to support.

Ah—the life of a development engineer.

Index

A

Y